Computer Communications and Networks

Series editor
A.J. Sammes
Centre for Forensic Computing
Cranfield University, Shrivenham campus
Swindon, UK

The **Computer Communications and Networks** series is a range of textbooks, monographs and handbooks. It sets out to provide students, researchers, and non-specialists alike with a sure grounding in current knowledge, together with comprehensible access to the latest developments in computer communications and networking.

Emphasis is placed on clear and explanatory styles that support a tutorial approach, so that even the most complex of topics is presented in a lucid and intelligible manner.

More information about this series at http://www.springer.com/series/4198

Veljko Milutinović • Jakob Salom
Nemanja Trifunovic • Roberto Giorgi

Guide to DataFlow Supercomputing

Basic Concepts, Case Studies, and a Detailed Example

Springer

Veljko Milutinović
School of Electrical Engineering
University of Belgrade
Serbia

Nemanja Trifunovic
Maxeler Technologies Inc.
Palo Alto, CA, USA

Jakob Salom
MISANU
Belgrade, Serbia

Roberto Giorgi
University of Siena
Italy

ISSN 1617-7975   ISSN 2197-8433   (electronic)
Computer Communications and Networks
ISBN 978-3-319-36758-3   ISBN 978-3-319-16229-4   (eBook)
DOI 10.1007/978-3-319-16229-4

Printed on acid-free paper

Springer International Publishing AG Switzerland is part of Springer Science+Business Media (www.springer.com)

Preface

This book is meant to support teaching of the DataFlow subject. University courses fully or partially dedicated to the DataFlow subject are important for the following reason: More and more Big Data are present in all kinds of research or commercial challenges. Consequently, the DataFlow paradigm is getting importance, since it has been proven that it is the most suitable computing paradigm for Big Data. It offers superior speedups (depending on the application, from about 20 to about 200, even 2000 in some isolated cases) as well as power savings (typically about 20 times); it brings size reduction, too. A recent study by researchers of the Tsinghua University in China reveals that for Shallow Water Weather Forecast (a Big Data problem), on the 1U level, compared to Tianhe-2 (at the time of writing of this book, rated #1 on the Top 500 SuperComputers list, which compares SuperComputers based on Linpack, a small data benchmark), Maxeler (a DataFlow machine) demonstrates the speedup of 14. With all the above in mind, the book is divided into four chapters: The first one is of an introductory nature. The second gives an overview of the related research. The third represents a case study. The fourth one is oriented to the ease of use and covers the issues of importance for WebIDE (a web-based integrated development environment). The work on this book was partially supported by the MISANU project #44006.

Belgrade, Serbia   Veljko Milutinović
Belgrade, Serbia   Jakob Salom
Palo Alto, CA, USA   Nemanja Trifunovic
Siena, Italy   Roberto Giorgi

v

Contents

Chapter 1
The DataFlow Paradigm

1.1 Introduction

Until a few years ago, one could say that all computers around us belong to the category of control flow computers. In the case of control flow computers, one writes a program to control the flow of data through the hardware. No matter how fast the today's control flow computers are and no matter how parallel they can be, the execution process is essentially slow, especially due to the continuous push-pull of data through the memory hierarchy and the synchronization across multiple cooperating threads. Also instructions have to be fetched, decoded, and executed. During the execution phase, the computer has to compute the addresses of data to be fetched, to fetch the data, to compute and store the result, etc. All these operations take time and the overall process can be extremely time-consuming.

Since the beginning of the 1970s, researchers had proposed DataFlow computers [Dennis74], but only recently they emerged on the market as a viable alternative for general-purpose computing. In the case of a DataFlow computer, one still writes a program, but not a program to control the flow of data through the hardware; instead, one writes a program that configures the hardware (in space), so, when data comes to its input ports, it just gets flown through the configured hardware, and the result is generated one, two, or even three orders of magnitude faster. How much faster depends on the characteristics of the application. The DataFlow is not driven by a program, but by a voltage difference between the input and the output of the hardware; consequently, the process can be extremely fast but also extremely power efficient and extremely small in physical size.

Many see the DataFlow approach as the most effective way to achieve the exascale speeds for big data applications [Maxeler2012]. However, few understand that all the potentials of DataFlow can be fully achieved only if the algorithms related to exascale applications are properly modified. Therefore, this book concentrates not only on how to design and program a DataFlow computer, but also on how to modify the existing algorithms for the best utilization of the DataFlow potentials.

© Springer International Publishing Switzerland 2015
V. Milutinović et al., *Guide to DataFlow Supercomputing*, Computer
Communications and Networks, DOI 10.1007/978-3-319-16229-4_1

The DataFlow approach represents a new old paradigm in computer design and programming. It is old, since the first ideas about DataFlow came all the way back in the 1960s and 1970s [Milutinovic88] by famous researchers like Jack Dennis (static DataFlow) and Arvind (dynamic DataFlow). It is new, because the enabler technology for implementation of the DataFlow hardware (FPGA) and software (OpenSPL) exists only for a relatively short time now. In other words, there was a relatively wide time gap between initial ideas and effective implementations, which is a characteristic of many important innovations.

As indicated above, the DataFlow paradigm, compared with the control flow paradigm, has three important advantages: speed, power, and size.

The speedups can go all the way up to 20, 200, or even 2,000 (application dependent). The power reduction is about 20 times (clock dependent). The size reduction is also about 20 times (paradigm dependent). These numbers are elaborated further in this book.

On one hand, comparing different architectures makes sense only for the same set of applications and the same set of data. On the other hand, comparing different architectures makes sense only if the design complexity and/or the purchase price is the same.

As far as the applications and data are concerned, the rest of this book deals only with big data applications and big data volumes.

As far as the design complexity and purchase price are concerned, one has to keep in mind the following: (a) If the design complexity is fixed, then one obtains all the three abovementioned advantages at the same time. (b) If the purchase price is fixed, one obtains the abovementioned advantages only close to all three at the same time, due to the fact that control flow computers are currently being produced in much larger quantities compared to DataFlow computers, thus lowering their production costs a great deal.

For all the above benefits to be achievable, certain conditions have to hold. Two are related to loop characteristics, two to application characteristics, and two to programmer characteristics.

Loop Characteristics

1. In essence, one can say,
   DataFlow technology migrates the execution of loops from software to hardware, which is an obvious method to make the loop execution faster.
   Ideally, the loop execution time is squeezed down to almost zero.
   In other words, for example, if a program takes 100 units of time to execute and 95 units of time is spent in loops,
   after the program acceleration based on the DataFlow approach,
   the program execution time is ideally 5 time units.
   Consequently, only the applications that spend minimum 95 % of time in loops could obtain a speedup of about 20 times.
   This is in accordance with the Amdahl's law.

2. The above discussion assumes that, after migration into hardware,
   the loop execution time approaches zero.
   The major question is, how much close to zero?
   The answer depends on the level of data reusability inside the migrated loop.
   The more reusability, the better the speedup.
   In reality, the speedups can be expected only
   if the level of reusability is higher than 3 (meaning reusing the same data 3 times).

Application Characteristics

1. The more the streaming in the application,
   the easier it is to overlap external communications (inherently slow)
   and internal processing (inherently fast).

2. Some applications do not tolerate even the smallest latency
   till the first partial result.
   Such applications may not be well suited for DataFlow implementation.

Programmer Characteristics

1. The programmer has to be familiar with the DataFlow programming paradigm,
   which is not difficult to comprehend,
   but takes time to learn it.

2. The programmer must possess an excellent understanding
   of the underlying algorithm
   and the underlying architecture,
   so he knows how to modify the algorithm and how to arrange input data
   for the best exploitation of the DataFlow concept.
   This can also be achieved if a domain expert is in the team.

Due to the fact that the programming effort is somewhat higher and that the compilation time is somewhat longer (both will be elaborated later), the DataFlow technology is best used for the so-called WORM (Write Once, Run Many) applications. In such applications, the cost of programming is amortized by the many runs of the compiled code and is not an economic issue anymore.

Modern supercomputers are carefully ranked using the Top500 Supercomputer List, initiated about 20 years ago by Jack Dongarra and others [Dongarra94]. In 2014, #1 on the list was the Chinese Tianhe 2. Before that are Cray Titan, IBM Sequoia, Japanese K, etc. There is no single DataFlow machine on that list. The question is how come is that possible if all the abovementioned DataFlow benefits really do exist?

The explanation is simple. Had the Top500 List used a big data benchmark rather than Linpack (which is not a big data benchmark) and had the list been concerned with all the three issues of importance (speed, power, and size), rather than with the speed alone (Top500) or speed and power (Green500), a DataFlow computer, like Maxeler, would be on the top of the list [STFC2014].

An elegant way to incorporate all the three issues (speed, power, and size) into an analysis is to compare machines (for big data benchmarks) by measuring how much speed can be obtained from a 1U box. If that is done, then, as a recent analysis demonstrates, Maxeler would definitely be on the top of the list [Flynn2013].

Generally, the DataFlow approach can be of the course structure type, as in the early work of Dennis and Arvind, or of the fine structure type, as in the case of Maxeler [Maxeler2015]. It looks like the fine structure approach is better suited for today's enabler technology: FPGA [Johnston2004].

In addition to Maxeler, some other companies are trying similar approaches, but Maxeler is by far the most successful on the markets of application giants like Schlumberger, JP Morgan, or Chicago Mercantile Exchange (CME), so the rest of this book uses examples exclusively based on the Maxeler approach.

In conclusion, the control flow paradigm implies that the compilation goes till the machine level code and that the execution process is performed at the machine code level. In the DataFlow paradigm, the compilation goes to the levels much below the machine code, i.e., to the levels of gates and wires, so the process is executed at the GTL (gate transfer level). As indicated before, this brings benefits (speed, power, and size) but also the challenges: a different programming paradigm and orientation to WORM applications.

The WORM applications are found in sciences (papers in geophysics report speedups of about 20–200), in banking (where speedups can go from about 200 to about 2,000), in image understanding, or in data mining from all kinds of sources, including also the social and sensor networks.

1.2   The Concept

Essentially, DataFlow computers are accelerators. One continues to run the old program on the host machine of the control flow type. When the time comes for a time-consuming loop to be executed (the assumption is that the loop satisfies the six aforementioned conditions), the execution is passed to the DataFlow accelerator that is connected to the host machine via a PCI Express bus (or for larger systems, via InfiniBand).

For the host and accelerator to integrate completely, one also has to download (at the host) the following additional pieces of system software: (a) the MaxelerOS (see Fig. 1.1), (b) the Maxeler compiler and the related run-time library (see Fig. 1.1), and (c) the Maxeler simulator.

Figure 1.1 assumes that two loops have been migrated from the host to the accelerator. This does not necessarily mean that only two loops existed in the host application; this means that the number of loops in the host application could have been higher, but only two loops satisfied the conditions for migration into the accelerator.

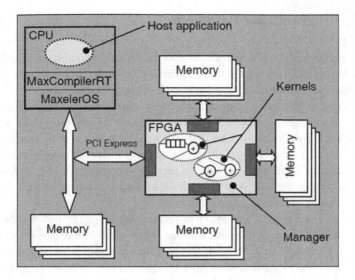

Fig. 1.1 Generic acceleration architecture (Courtesy of Maxeler Technologies)

For each loop, one has to write two program units: (a) kernel and (b) kernel test. This means, if n loops are migrated, one has to write $2n$ new program units.

Also, no matter how many loops are migrated, one has to write another three new program units from scratch: (a) manager, (b) simulator builder, and (c) hardware builder. A detailed example is shown in Chap. 3.

Therefore, the total of $2n + 3$ new program units have to be written. All these programs are written in Maxeler Java, which is a superset of Java, with dozens of new functionalities (built-in classes) added on the top of the classical Java. Consequently, Maxeler Java is a new DSL (domain-specific language). Recently, Maxeler Java emerged into a new and a more elaborated language environment called OpenSPL (Open Spatial Programming Language).

The host can be a simple PC or a sophisticated supercomputer. The accelerator itself can be a PCI Express board; a 1U box; a workstation with several 1U boxes; a rack with 10, 20, or 40 1U boxes; or a train of racks.

In general, there is a 1:1 correspondence between the number of loops migrated into the accelerator and the number of kernels that one has to write from scratch. However, sometimes we have more kernels than the migrated loops, or the opposite.

We have more kernels if one loop is described with more kernels (i.e., if one uses the "divide and conquer" strategy). We have fewer kernels than the migrated loops if one kernel can be used for the execution of two different loops in two different parts of the host application (i.e., if one uses the "time sharing" strategy).

1.3   The Programming Paradigm

Figure 1.2 represents the first introduction to the OpenSPL language Maxeler Java (MaxJ), using an example related to the computation of the moving average, which is an essential element of all algorithms based on convolution.

The MaxJ programming language includes two types of variables: (a) the standard Java variables that will be referred to here as *software Java variables* and (b) the DataFlow Java variables that will here be referred to as *hardware Java variables*. The software Java variables are there to instruct the compiler what to do, so they disappear after the compilation process is completed. The hardware Java variables are there to actually flow through the hardware, so they produce the results after the DataFlow process is completed.

As indicated in Fig. 1.2, the MaxJ programming language follows the syntax of the Java language, except for the add-ons. For example, a hardware variable is written under the quotes (e.g., "x") while it is in the environment outside the Maxeler system. Once it enters the Maxeler system, the quotes are eliminated (e.g., x, without quotes). The above described is well seen in Fig. 1.2, together with the convention used to define the appropriate hardware variables (HWVar or DFEVar), as well as to define the floating-point precision, exponent (8 bits) and mantissa (including sign, 24 bits). All other hardware variable types are defined similarly (elaborated later).

The Maxeler compiler produces a graph, as given in the right-hand part of Fig. 1.2. The graph can be presented in a graphical form (as indicated in Fig. 1.2), or it can be described in VHDL (which is of importance for further processing). Further processing, from the graph level (left by the Maxeler compiler) till the binary level (needed for configuration of the FPGA circuitry), is done using the synthesis tool of the manufacturer of the FPGA circuitry used to implement the DataFlow system

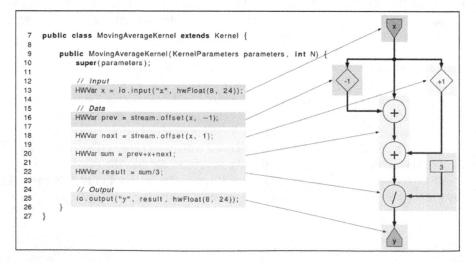

```
 7   public class MovingAverageKernel extends Kernel {
 8
 9       public MovingAverageKernel(KernelParameters parameters, int N) {
10           super(parameters);
11
12           // Input
13           HWVar x = io.input("x", hwFloat(8, 24));
14
15           // Data
16           HWVar prev = stream.offset(x, -1);
17
18           HWVar next = stream.offset(x, 1);
19
20           HWVar sum = prev+x+next;
21
22           HWVar result = sum/3;
23
24           // Output
25           io.output("y", result, hwFloat(8, 24));
26       }
27   }
```

Fig. 1.2 MaxJ (Maxeler Java), the code and the related graph (Courtesy of Maxeler Technologies)

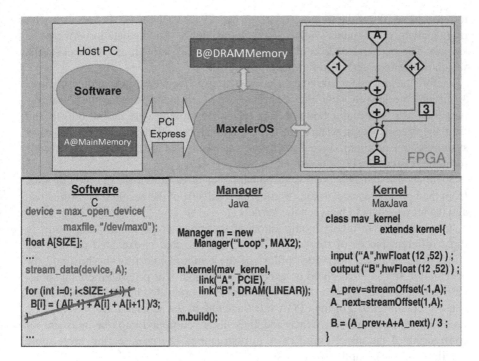

Fig. 1.3 Hardware/software relationship in a DataFlow system, using Maxeler as an example (Courtesy of Maxeler Technologies)

(typically Altera or Xilinx), which is elaborated later on. Figure 1.3 describes the relationship between hardware and software components in a DataFlow system, again using Maxeler as an example.

The "for" loop from the main program (originally written to run on the host) in Fig. 1.3 is assumed to be the loop to be migrated from the host to the accelerator. For that to happen, the program segment implementing the loop has to be deleted (the red line in Fig. 1.3), the input data for the loop have to be moved from the host memory area A to the accelerator internal memory, and after the results related to this loop are generated, they (the results) have to be moved from the accelerator internal memory to a memory area B on the accelerator board/box for further processing (and eventually, at the end of the entire loop-related process, sent back to the host). The data movement is realized by the "stream_data(device, A);" construct, which implies that the "device" has to be defined first ("device = max_open_device(maxfile, "/dev/max0");"). As indicated in Fig. 1.3, all these changes happen in the host code.

It is the manager code that is responsible for accepting data on the accelerator side, which (the acceptance of data) is implemented via "link("A", PCIe)". It is also the manager code that is responsible for moving the partial results ("link("B", DRAM(LINEAR)")") to the memory area B in the accelerator (and eventually, for moving the results of the looping-related process back to the host). In the cases with more kernels, the manager code is also responsible for moving data in between the kernels. The manager code is also enlisted in Fig. 1.3, as well as the kernel code, which is here repeated from Fig. 1.2.

Before the processing starts, data has to be moved from the external memory area A through a relatively slow PCI Express bus, which means that, in the DataFlow paradigm also, the major bottleneck is related to moving data in and out, to and from the system. Moving data to the temporary memory area B inside the accelerator system is not a problem, since it can happen at speeds that are an order of magnitude (or more) faster than PCIe transfers.

Overall, the manager code is responsible for three activities: (a) moving data from the host to the accelerator, (b) moving results from the accelerator to the host, and (c) moving data in between the kernels.

Consequently, in theory, compilation of the manager program could result in three different pieces of the "object" code: (a) one that is executed on the host CPU, to serve as the host-side "pillar" of the bridge connecting the host and the accelerator, (b) one that could be executed on a small (for programmers invisible) CPU on the accelerator side, and (c) one that is incorporated into the .max file that is used for configuring of the FPGA infrastructure of the accelerator.

As far as the kernels are concerned, they all are compiled into one .max file for configuring of the underlying FPGA infrastructure. The next section gives more details about the compilation process.

1.4   The Compilation

Figure 1.4 illustrates the compilation process and helps understand the values of the time constants related to the compilation process.

The compilation process is entered with (a) one or more kernel programs (.java files), (b) one manager program (also a .java file), and (c) the host program (e.g., a .c file), as indicated in Fig. 1.4. With the help of *MaxelerOS* (a run-time system provided with the Maxeler board), for building hardware, the compiled kernel and manager code are combined, and the execution graph is formed. In Fig. 1.4, the execution graph is depicted with the horizontal line at the top of the "Hardware Build" block. The Maxeler simulator works on the execution graph level, so that level will be denoted here as the simulator level.

Before generating the FPGA-related files (.max), a user typically tests the correctness of the code by using the simulator level framework. To move from the simulator level to the .max level, one has to use the tools from the manufacturer of the underlying FPGA technology (to generate the production code for FPGAs). Once the .max file is formed, it is linked with the compiled .c code and the appropriate routines from the MaxCompilerRT (RT stands for run time). The final product of the compilation and linking is the executable for the control flow host ("Application" in Fig. 1.4).

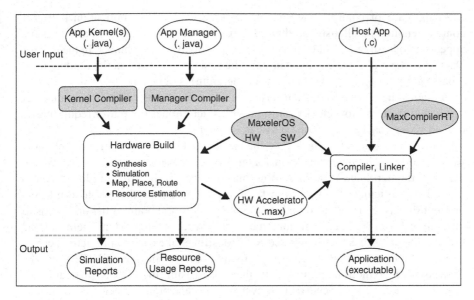

Fig. 1.4 Details of the compilation process (Courtesy of Maxeler Technologies)

When the execution starts on the host machine, a communication is established with the accelerator, and the accelerator is "asked" if it is configured with the related .max file. If it is not, then the accelerator is configured, by moving the .max file from the host to the accelerator. After the configuration is completed, the big data stream has to be activated, and the execution starts. On the next rerun, no initial configuring is needed, unless another user was using the hardware in the meantime. Of course, if another user was using just a part of the hardware in the meantime and did not damage the configuration used by the initially mentioned program, then no reconfiguration is needed.

What are the time constants involved in the process?

Compilation from the level of kernels to the .max level could take several hours. Compilation from the level of kernels to the graph level could take several minutes. Loading of the .max file and configuring of the FPGA hardware could take several seconds. Starting a big data stream may take a few milliseconds. Starting the execution of the compiled code may take several microseconds. Of course, the values of all these constants are likely to change over time, as the technology changes.

1.5 Comparisons

A crucial question is how the two concepts compare: control flow and DataFlow. Both concepts subdivide into two major categories.

The control flow concept subdivides into multi-core and many core. The DataFlow concept subdivides into coarse grain and fine grain.

In the case of control flow, the major issue is that the application software contains enough parallelism, so that all cores can be kept busy, and partial results appear periodically in small (multi cores) or big (many cores) chunks. In the case of DataFlow, the major issue is the delay till the first partial result on the module level (coarse grain) or the loop iteration level (fine grain) is obtained.

As far as the clock speed is concerned, it is typical that the control flow clock is faster, and the DataFlow clock is slower, which determines the power requirements of the two approaches, as it will be discussed later.

Occasionally the researchers joke that using an N-core system is like plowing with N horses. The major question for multi-core systems (e.g., Intel) is: Which way will the horses go? For them to go into the desired direction, a plow driver is needed. We articulate our wishes in the form of a program that we load into the brain of the driver. In order to understand our wishes, the brain of the driver has to be equipped with von Neumann constructs, and in order to understand them fast, the brain of the driver has to be equipped with constructs like caches, predictors, etc.

In the same anecdotic style, using a many-core system (e.g., NVIDIA) is like plowing with N thousand chicken. In that case (using the CUDA programming model), more drivers are needed in the system, and each brain of each chicken is also equipped with von Neumann constructs. In the more recent rCUDA programming model (remote CUDA), in addition to more drivers, one dispatcher is needed, too.

The question now is what one plows with in the case that a DataFlow system is used? The answer is: with N million ants. Each ant has a backpack for a portion of big data. The idea is that, while traversing the plow field, the ants generate the end result. For this to happen, one first has to configure the field (to write kernel programs) and to load the backpacks of the ants (to move data into the accelerator). What motivates the ants to move into the right direction, in conditions when no control flow program exists? The data is moved by the voltage difference between the input and the output of each gate in the FPGA structure.

Why is the DataFlow paradigm so fast? Because it compiles down to levels much below the machine code level; it compiles down to RTL (register transfer level), GTL (gate transfer level), and TTL (transistor transfer level), as indicated in Fig. 1.5;

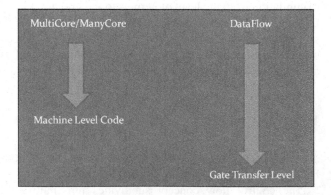

Fig. 1.5 The DataFlow paradigm compiles to lower hardware levels

MultiCore/ManyCore   DataFlow

Machine Level Code

Gate Transfer Level

Fig. 1.6 The DataFlow paradigm uses a slower clock (200 MHz vs. 4 GHz)

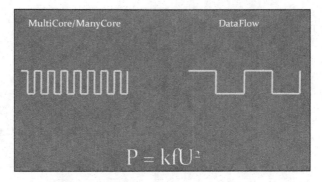

moreover, it avoids the continuous push-pull of data through the memory hierarchy (typical of the control flow-based processors). Another way to look at the issue is by stating that the DataFlow paradigm uses bit serial arithmetic and internal arithmetic pipelines, which enable more computation to be done per time unit.

Why is the DataFlow paradigm so much more power efficient? Because the power dissipation could be calculated using the following formula:

$$P = f\left(f, U^2\right) = kfU^2$$

where U is the voltage applied, f is the frequency applied, and k is a constant. No matter if control flow or DataFlow is being used, U is the same, and the constant k is the same for the same implementational technology. What differs is the frequency f. In the case of modern control flow multi-core implementations, the frequencies go up to about 4 GHz (e.g., Intel). In the case of modern control flow many-core implementations, frequencies are lower but still relatively high. In the case of FPGA-based DataFlow implementations, frequencies are as low as about 200 MHz (e.g., Maxeler Technologies). If we divide 4 GHz by 200 MHz, we get 20, which is the power savings ratio most frequently mentioned in the literature. Figure 1.6 is meant to portray these facts mnemonically.

Finally, why is the size of a DataFlow implementation so much smaller? It is so because the control flow paradigm is based on the von Neumann architecture in which only about 5 % of the chip area is dedicated to the ALU area (the rest being used for nonfunctional silicon, such as caches, branch predictor, out-of-order execution, etc.), while the DataFlow paradigm is best implemented on the top of FPGA architectures in which more than 95 % of the chip area is dedicated to ALU-type functionalities. The ratio of "more than 95 %" and "about 5 %" is about 20, which is the number most frequently quoted in the literature (see Fig. 1.7 for a mnemonic explanation).

If one likes to preserve the 20:1 ratio in size even after packaging the chips, one has to place (into the package) very small coolers and very small fans. Consequently, the DataFlow implementations are noisy. The noise comes for two different reasons: (a) smaller fans are noisier and (b) since coolers are smaller, the fans must rotate faster, and fast fan rotation is the second major source of fan noise.

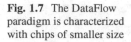

Fig. 1.7 The DataFlow paradigm is characterized with chips of smaller size

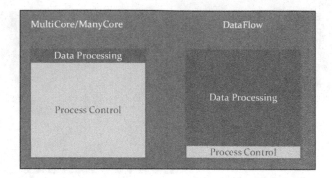

The programming effort is not higher. The number of new programs to write is higher $(2n + 3)$, but they are straightforward to write, and one does not have to worry about issues like the influence of caches, TLBs, predictors, or memory consistency. However, it does not mean that one does not have to worry about memory hierarchy. Actually, the memory hierarchy does exist in DataFlow systems, typically two levels inside the FPGA chips and another two levels on the accelerator board outside the FPGA chips. That memory hierarchy is to be worried about at programming time, when kernels are written (not at run time, like in the case of control flow machines).

The debugging effort might be higher, since more lines of code may mean more opportunity for a bug to happen. That is why a strong system support had to be developed in software packages supporting the OpenSPL concept [Maxeler2014].

Finally, the needed compilation efforts deserve special attention. As indicated before, compilation takes time, and therefore, the preferred applications are the ones that are compiled once and run relatively large number of times.

1.6   DataFlow Hardware (A Maxeler Example)

Maxeler provides ready-to-use machines that exploit the DataFlow paradigm. The current Maxeler offer includes three different options: (a) the C option representing a full-blown computer (C = computer), (b) the X option representing an accelerator (X = accelerator), and (c) the N option for low-latency LAN applications (N = network).

The 1U of the C option includes four DFEs (DataFlow engines), plus a control flow host. This option provides a full-blown computer, rather than just an accelerator. The reason for adopting the C option may be in the fact that the user computer center does not own an InfiniBand bus, which is needed (could be in addition to the PCIe bus) for bigger Maxeler systems.

The X option includes no control flow host, which leaves space for 8 (not 6) DFEs – all that for approximately the same purchase price, which makes the accelerator option much more desirable if the user does own the InfiniBand bus.

Finally, N option minimizes the latency (which is the Achilles' heel of DataFlow supercomputing) and can be used in latency-intolerant applications, like trading or financial analytics, where trading data and data to be analyzed arrive via a network. It is obvious, in trading applications, even if one uses the same buy-sell algorithm as the competition, one has an obvious advantage if supported by high speeds that the DataFlow paradigm offers.

In the latest production of Maxeler, the amount of onboard memory is 768 GB (and approaching over 1 TB). This means, the technology is best used if the big data can be partitioned into chunks of 768 GB (or 1 TB in near future).

1.7 Application Types

We can identify at least three major groups of applications for DataFlow super-computers: (a) coarse grained, stateful, typically used in business; (b) fine grained, transactional, with a shared database, typically used for data mining; and (c) fine grained, stateless, transactional, typically used in scientific applications, in physics, chemistry, etc.

In the first case (coarse grained, stateful), the CPU needs a DFE for minutes or even hours, and the DFE is typically engaged after an interrupt request. Examples of this approach can be found with JP Morgan or CME (Chicago Mercantile Exchange), for jobs related to computation of credit derivatives or for online trading. In the case of online trading (OLT) and high-frequency trading (HFT), the computational infrastructure has to be equipped also with a connection to an atomic clock and an Ethernet connection to trading data.

In the second case (fine grained, transactional), CPU utilizes a DFE for milliseconds to seconds; each short computation is typically performed on the data fetched from a shared memory. Examples of this approach could be found in data mining from social networks, like Facebook or Twitter.

This case is also typical of applications in which data mining is performed on the top of data related to credit card transactions, or if Mind Genomics [Moskowitz2014] analysis is performed on the top of marketing data.

In the third case (fine grained, stateless), CPU also utilizes a DFE for only milliseconds to seconds, performing many short computations with the contents of a private database. Examples of this approach could be found with Schlumberger (oil and gas) for deciding where to drill the ocean floor or with Earth Sciences (Weather Forecast) for a more accurate weather prediction. In the oil and gas industry, when using DFEs to decide where to drill for oil and/or gas, seismic vessels are used, as indicated in Figs. 1.8a and 1.8b.

Seismic vessel periodically emits a water blast into the ocean floor. The blast partially reflects off the ocean floor and partially penetrates into the floor. Under the ocean floor, the same happens at each and every seismic layer underneath: a part of the wave reflects and the rest penetrates. Once the waves reach the gas and/or oil

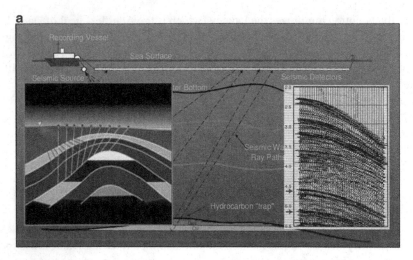

Fig. 1.8a Deciding where to drill for oil and/or gas. X axis refers to horizontal distance; Y axis refers to vertical depth. For more details see Sect. 2.2.6 (Courtesy of Schlumberger)

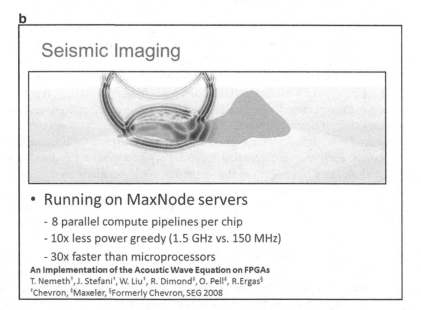

Fig. 1.8b Deciding where to drill for oil and/or gas. The wave is partially bouncing back and partially penetrating towards the lower seismic layers (Courtesy of Maxeler Technologies)

layers, the same happens, except that the penetrated waves are not of interest any further, only the reflected waves. They are collected by a carpet of sensors being dragged by the seismic vessel. The data collected by the carpet of sensors has to be processed by a set of partial differential equations. The result of the processing is the major parameter in the decision-making process at the given geographical location: to drill or not to drill?

The main question is: Where does the processing of partial differential equations take place? The first guess of a typical reader is that the processing takes place at the seismic vessel, which is not the case, due to the power restrictions. The vessel is not able to create all the energy needed for the processing. Consequently, data is sent to a mainland supercomputer center, not via wireless, since this is big data and the wireless approach would take too much time, instead by moving the storage devices (i.e., hard disks) via a helicopter, which takes orders of magnitude less time to transmit!

A paper by Nemeth et al. [Nemeth2008] proves, for the abovementioned application, that an implementation of a DataFlow machine is 30 times faster compared to an implementation of an Intel machine of approximately the same market value. At the same time, the DataFlow machine was working with a clock of 150 MHz, while the Intel machine was working with a clock of 1.5 GHz. The latter means that the DataFlow machine power consumption was 10 times smaller. Consequently, the speed/power factor is about 300 times better in favor of the DataFlow machine.

While power at a given large scale may become a scarcely sustainable cost, it has to be considered a critical factor also for medium-size data centers, where upgrading the power infrastructure may imply an exponential cost increase due to a complete reengineering, new contracts with the power provider, and new powerlines.

Since the speed and power could be traded, in theory, one can arrange that the speed of the DataFlow and the control flow machine be the same, in which case the power consumption of the DataFlow machine can be 300 times smaller, and the above-described processing can be done at the seismic vessel.

1.8   Application Examples

In the domain of surface reflection analysis, Maxeler sources demonstrate that a speedup of 230 times is possible for land cases (a MAX2 card compared with a CPU) and a speedup of 190 times in the marine case. Figure 1.9 shows that the final results of two competing programming paradigms produce almost identical output of a complex processing (in this case CRS – Common Reflection Surface – processing).

In the domain of seismic trace applications, an example from Italy (ENI-AGIP) demonstrates the fact that 100 MAX2 cards were showing approximately the same performance as the system with 21800 CPUs, for the conjugate gradient application used for seismic tracing. Figure 1.10 shows that the final results of the two alternative approaches are approximately the same [WANG2010].

In the domain of angle gatherers applications, a speedup of 48 times was obtained at the [SEG2008]. Figure 1.11 shows the formula used in this application, which boils down to the computation of the correlation function for a number of different angles.

Fig. 1.9 The common
reflection surface (CRS)
analysis – end results based
on the two competing
programming paradigms are
almost identical, although the
DataFlow-based results were
generated either 230 or 190
times faster. For more details,
see Sect. 2.2.6.1 (Courtesy of
Maxeler Technologies)

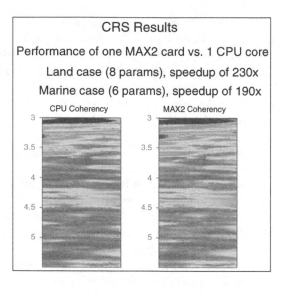

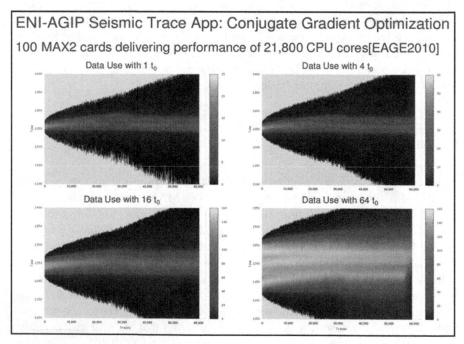

Fig. 1.10 The conjugate gradient application; data use cases with 1 t_0 (essential parameter for CGA applications), 4 t_0, 16 t_0, and 64 t_0 (Courtesy of Maxeler Technologies)

48x Speedup of Angle Gathers
with Stanford Center for Earth and Environmental Sciences [*)]

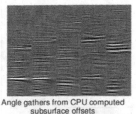

Angle gathers from CPU computed
subsurface offsets

- Can be dominant cost in shot profile migration
- Cross-correlating two fields by various shifts:

$$I(h,x,z) = \sum_{s}\sum_{w} S(x-h,z,w,s) \cdot G^{*}(x+h,z,w,s)$$

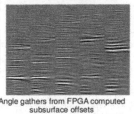

Angle gathers from FPGA computed
subsurface offsets

**SPEEDUP RESULTS FROM CUSTOM
TRACE MEMORY SYSTEM:**
- **Trace = Unit of Transfer**
- **Buffers Prefetch Right Traces in Advance**

Fig. 1.11 Data use cases for various angle gathers in 3D seismic imaging (Courtesy of Maxeler Technologies)

1.9   Acceleration Is Hard

In order to generate the maximal possible acceleration for a given algorithm, one has typically to work hard in the following four domains:

(a) Appropriate algorithmic modifications
(b) Exploiting pipelining concepts
(c) Appropriate input data choreography (i.e., reorganization of the input data)
(d) Exploiting FPGA possibility for arbitrary floating-point precision

As far as algorithmic modifications are concerned, one has to adapt the order of operations to the internal architecture of the DataFlow supercomputer, trying to avoid any negative impact on the integrity of the algorithm. For example, instead of first summing up elements of a vector and then multiplying the sum with a constant, one can first multiply the vector elements with the constant and then sum up the partial products, to produce the same result. Figure 1.12 shows how Sasa Stojanovic [Stojanovic2015] adapted the Gross-Pitaevskii algorithm for implementation on a DataFlow machine.

As far as pipeline utilization is concerned, one has to adapt the algorithmic steps to the structure of the internal pipeline of the utilized DataFlow processor. For example, instead of a step-by-step progress through the pipeline, one can first fill out the entire pipeline and then proceed step by step. Figure 1.13 shows how in the

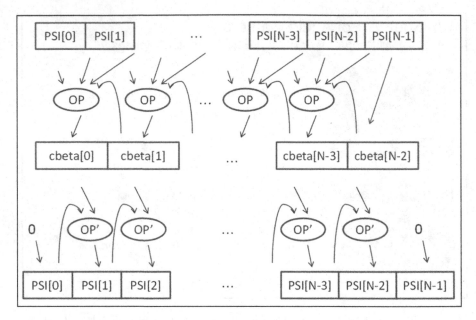

Fig. 1.12 Modifications of the Gross-Pitaevskii algorithm for DataFlow implementation [Stojanovic2013]

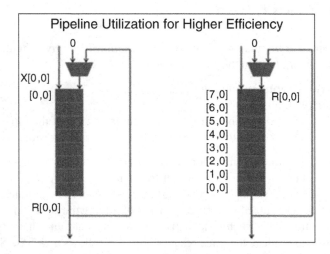

Fig. 1.13 Alternative utilizations of the internal pipeline in a dataflow machine (bad, *left*; good, *right*). Bad, because the pipeline is underutilized and delivers data every eighth cycle; Good, because the pipeline delivers data every cycle [Stojanovic2013]

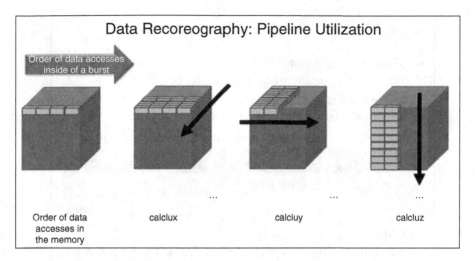

Fig. 1.14 One input data choreography tuned to a DataFlow implementation based on Xilinx Virtex-6 [Stojanovic2013]

same example the Gross-Pitaevskii algorithm was changed for better utilization of an internal pipeline of a DataFlow representation of the algorithm. Bad means filling the pipeline one by one and good means filling the pipeline all at once.

As far as input data choreography is concerned, one has to adapt the input data chunks to the size of the internal structure, on each and every architectural level. For example, if the internal DataFlow architecture is based on Virtex-6 Xilinx components, then it is best to enter the data six by six (the best suited combination for Virtex-6 Xilinx architecture). If the size of the internal memory is 768 GB (8 times 96 GB), then the problem has to be decomposed into data chunks smaller than 768 GB. Figure 1.14 shows how in the mentioned example the input data was organized for the implementation on a DataFlow machine based on Virtex-6 [Stojanovic2015]. The *calclux*, *calcluy*, and *calcluz* are procedures related to the three different dimensions of the Gross-Pitaevskii algorithm.

As far as the optimal selection of the floating-point or fixed-point precision level is concerned, in control flow machines, the word size is fixed to 32 or 64 or a multiple there-off bits; in DataFlow machines, one can organize own fixed-point and floating-point structures, with a minimal number of bits needed for the mantissa and the exponent in the given application.

Once the precision level is decreased, the saved hardware can be reinvested into a higher speedup. In the case of Maxeler, a special 24-bit format is introduced, compatible with the structure of the FPGA chips used for hardware implementation, which is a feature that provides ground for enormous hardware savings, as indicated in Fig. 1.15. The Add and Multiply lines are related to add and multiply on the FPGA level.

Fixed Point: Savings Reinvestable

- Consider fixed point
  compared to single precision floating point
- If the range is tightly confined,
  one could use *24-bit* fixed point
- If data has a wider range, may need *32-bit* fixed point

	hwFloat(8,24)	hwFix(24,...)	hwFix(32,...)
Add	500 LUTs	24 LUTs	32 LUTs
Multiply	2 DSPs	2 DSPs	4 DSPs

- Arithmetic is not 100% of the chip.
  In practice, often ~5x performance boost from fixed point.

Fig. 1.15 Savings when fixed-point arithmetic is used (Courtesy of Maxeler Technologies)

1.10 Open Research Problems

The following research problems are currently seen as the major ones on the way towards a wider adopting of DataFlow computers: (a) revisiting the supercomputer ranking systems, (b) revisiting the major supercomputing algorithms, (c) creating new architectures for better match between the FPGA architectures and the optimizing compilers for DataFlow processing, (d) improving reliability and fault tolerance of the infrastructure level, (e) solving the memory bandwidth problems, and (f) making a more efficient system software and tools that enable DataFlow programmers to be more effective.

As far as the ranking of supercomputers is concerned, the major problem today is the fact that the benchmarks used are not oriented to big data. For example, the Top500 List uses the benchmark called Linpack, which is a toy benchmark, rather than a big data benchmark. Also, the top supercomputer lists are typically concerned with speed alone or the ratio of speed and power, and not with how much speed one can obtain from a 1U box, which is a performance measure that takes into consideration all the major issues: speed, power, and size.

Consequently, on one hand, DataFlow supercomputers are never on the top of official lists, and on the other hand, supercomputer centers claim that, for their applications, a DataFlow supercomputer outperforms the machines on the top of the official lists [STFC2014] and that fact gets recognized ways away from research communities, as indicated in Figs. 1.16 and 1.17.

As far as the revisiting of algorithms is concerned, all algorithms of interest have to be revisited for the optimal implementation in a DataFlow environment. This

Fig. 1.16 Angela Merkel and David Cameron with the CEO of a DataFlow company

type of activity is difficult to optimize at the compiler level. Therefore, it has to be optimized by talented DataFlow programmers and made available to the community.

One of the major problems of the DataFlow programming is also on the psychological side: often a novice programmer may not reach the speedup potentials of a given DataFlow machine. The problem arises when such a novice goes around telling that the DataFlow approach is not a good one, instead of admitting that being a novice he/she still has not acquired the knowledge to appreciate that approach and to get sizable results out of it.

Direct migration of algorithms from control flow to DataFlow would typically result in a speedup which is not a substantial one. A better speedup is obtained after appropriate algorithmic changes, an even better one after appropriate utilization of internal pipelines, and the best possible speedup is obtained after appropriate input data re-choreography. Additionally, if the final application is of the floating-point or fixed-point type that can tolerate a lower precision, then obtained speedups can be enormous.

Consequently, one possible research avenue is to work on the development of new algorithms or modifications of the existing ones, for the best utilization of the DataFlow concept.

Another possible research avenue is to rank the algorithms separately for the control flow and the DataFlow environments. It may happen, if a given application

> ## World's Most Efficient DataFlow Supercomputer at STFC Daresbury Laboratory to Drive UK Science and Innovation
>
> The Science and Technology Facilities Council (STFC) and Maxeler are collaborating in a project funded by the UK Department of Business Innovation and Skills to install the next generation of supercomputing technology in a new facility at the DaresburyLaboratory focusing on energy efficient computing research. *The supercomputer will offer orders of magnitude improvement in performance and efficiency.* The new MPC-X supercomputer will be available in Summer 2014 and will allow UK industry and academia to develop products and services based on MPC data analytics engines for *applications domains such as medical imaging and healthcare data analytics, manufacturing, industrial microscopy, large scale simulations, security, real-time operations risk, and media/entertainment.*
>
> The dataflow supercomputer will feature Maxeler developed MPC-X nodes capable of an equivalent 8.52TFLOPs per 1U and 8.97 GFLOPs/Watt, *a performance per Watt that tops the Green500 today.* MPC-X nodes build on the previous generation technology from Maxeler deployed at JP Morgan where real-time risk computation*equivalent to 12000 x86 cores was achieved in 40U of dataflow engines.* For the full story please visit our website at:
>
> http://www.maxeler.com/stfc-dataflow-supercomputer/?utm_source=Commercial+List&utm_campaign=862e1b9d0e-CommercialFebruary2014_Mailer&utm_medium=email&utm_term=0_ece0f8fd2e-862e1b9d0e-336335821.
>
> 79

Fig. 1.17   A report from STFC (Courtesy of Maxeler Technologies)

can be realized with a number of different algorithms, that the algorithm which is the worst one in the control flow environment is the best one in the DataFlow environment, and vice versa.

As far as new architectures are concerned for better match between the compiler and the architecture, one has to have in mind the notions of Feynman [Feynman96] saying that the major advantage of DataFlow computing is in the fact that communication delays can be made zero or almost zero. This means that the major task of a DataFlow compiler is to generate a planar graph with zero or almost zero communication lengths. Assuming that a smart enough compiler is able to generate such a graph, the question is if the architecture can host such a graph without the needs to create deviations due to the fact that there is no good match between the compiler output and the targeted computer architecture. Current FPGA structures are matrix organized, while the graphs generated by compilers are typically treelike. Consequently, research is needed to create FPGA structures that would represent a better match for the compiler output. Figure 1.18 shows a typical FPGA topology and a typical DataFlow graph topology.

As far as reliability and fault tolerance are concerned, the major problem is how to create mechanisms that could reconfigure at run time, without recompilation, if a hardware bug gets generated during the system exploitation time. This issue was one of the project goals of the FP7 project BALCON [Milutinovic2014].

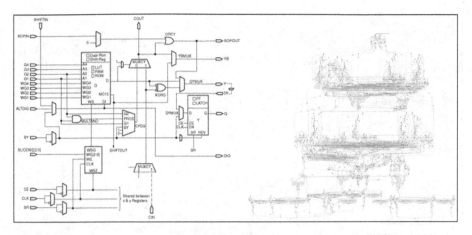

Fig. 1.18 Topologies of a typical (a matrix) FPGA chip and a typical dataflow graph (a set of trees) with about 5.000 computational nodes

As far as the memory bandwidth issues are concerned, current in-system bus speeds of a DataFlow machine well exceed the speed of 1 TB/s. However, the speed of external busses is about 100 Gb/s. It is difficult and expensive to reach over 100 Gb/s. Also, for the internal memory capacity, it is currently difficult and expensive to reach over 1 TB. Since the DataFlow paradigm is mostly meant for big data, the above means that big data has to arrive in streams and to be locally stored in chunks. The bigger the data chinks, the better the performance. Therefore, lots of research is needed for implementation of this goal.

As far as the system software and intelligent tools are concerned, the major challenge is to create the compiler that generates planar graphs with zero delays and also to create a WebIDE environment with examples and case studies. One such effort is available at http://webide.maxeler.com and will be elaborated later in this study. Another such effort is making an efficient SLiC (Simple Live CPU) Interface. See Fig. 1.19 for programming languages available at the SLiC level.

The interest in computing systems based on DataFlow principles is represented also by relevant investments both in Europe (e.g., the TERAFLUX project [Giorgi2014]) and in the USA (e.g., the DoE-funded X-Stack project [DOE2014]). The Maxeler approach is pushing these concepts further.

1.11   Enabler Mechanisms for DataFlow Supercomputing

For the DataFlow technology to become more acceptable, specific actions must be taken and specific mechanisms have to be established; some of them are in the techno-economic domain and some are in the techno-industrial domain.

In the techno-economic domain, the main issue is to make the technology accepted by potential users to which it can be of benefit. On the long run, the best

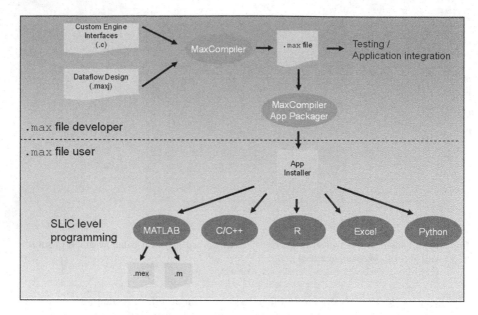

Fig. 1.19 Tools currently available from Maxeler (Courtesy of Maxeler Technologies)

way is to go through universities. That is why it is wise to give the system software and the system simulator to universities free of charge. Also, the education-oriented DataFlow machines should be provided to these universities free of profit, meaning that the purchase price should be equal to the price of the components used inside plus the cost of their assembly. On the short run, the best way goes through private industry eager to minimize costs and maximize profits. Of course, institutions financed from state budgets are the slowest in accepting new technologies, no matter how useful and profitable they can be.

In the techno-industrial domain, the major issues are (a) minimization, so the DataFlow concept can enter the cell phone market, for example, to accelerate a number of applications, (b) experimenting with new approaches to FPGA, and (c) enabling a variety of programming languages and application languages to enter the DataFlow environment.

As far as minimization is concerned, the major challenge is to port the DataFlow concept from its current 1U boxes or PCIe boards to chips, which is an important task in front of the semiconductor industry. A current interesting trend is represented by chips like the Xilinx Zynq, which integrate on the same chip a dual-core ARM A9 and programmable logic (FPGA).

Minimization is of importance not only for the commercial markets (like smartphones) but also for the defense markets (e.g., drone applications).

As far as new approaches to FPGA are concerned, the point is in the fact that current DataFlow technology works on 200 MHz, while some FPGA vendors have announced the FPGA chips on 2 GHz. This does not say that by simple FPGA substitution one gets a speedup of another order of magnitude – the whole "game" is much more complex than that. This only says that the advances in FPGA technology will not bring a slowdown of the DataFlow concept, so the future of the "game" is optimistic.

As far as enabling new languages and new applications is concerned, support is needed for absolutely all high-level programming languages (HLLs), meaning that the programs to be accelerated could be originally written in any programming language and on the top of any operating system (OS). The same should be the case for all major application packages and related application languages.

1.12 Getting Started

To make the presentation more concrete, this section is based on the Maxeler approach to DataFlow. Once the system software and the simulator are downloaded, a procedure has to be followed as in Fig. 1.20. Of course, if a WebIDE is used, as it will be indicated later, the steps from Fig. 1.20 become "invisible."

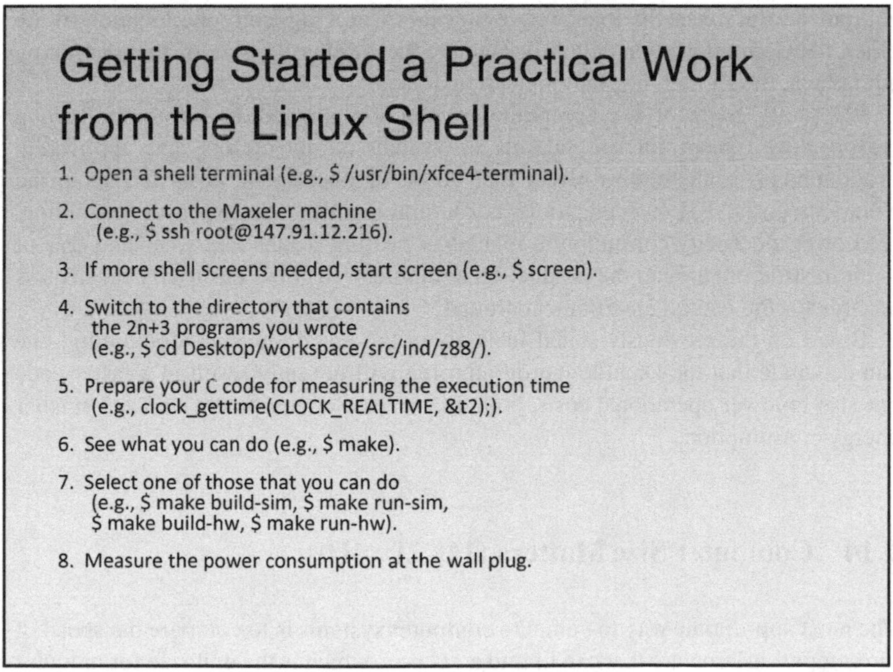

Getting Started a Practical Work from the Linux Shell

1. Open a shell terminal (e.g., $ /usr/bin/xfce4-terminal).

2. Connect to the Maxeler machine
   (e.g., $ ssh root@147.91.12.216).

3. If more shell screens needed, start screen (e.g., $ screen).

4. Switch to the directory that contains
   the 2n+3 programs you wrote
   (e.g., $ cd Desktop/workspace/src/ind/z88/).

5. Prepare your C code for measuring the execution time
   (e.g., clock_gettime(CLOCK_REALTIME, &t2);).

6. See what you can do (e.g., $ make).

7. Select one of those that you can do
   (e.g., $ make build-sim, $ make run-sim,
   $ make build-hw, $ make run-hw).

8. Measure the power consumption at the wall plug.

Fig. 1.20 Initialization steps from the Linux Shell (Courtesy of Maxeler Technologies)

All the explanations in Fig. 1.20 should be self-explanatory or intuitive, assuming that the reader has a basic understanding of the Linux OS. For more details about Linux, one is referred to existing textbooks, e.g., [Linux2000].

1.13   OPEX

Operational expenses (OPEX) are often times neglected in studies related to software costs. However, they represent an important fraction of the system cost. Fortunately for DataFlow supercomputing, the DataFlow approach is characterized with much lower operational costs, compared to the control flow approach. These facts are elaborated next.

Physical size of the computing equipment does matter, for the following reasons: (a) The space for computing equipment has to be rented, which can be a nontrivial cost; the smaller the size of the computing equipment, the smaller the rent. (b) Management of the rented space for the computing equipment also costs money (heating, cooling, cleaning, washing, etc.); the smaller the size of the computing equipment, the smaller the maintenance costs. (c) From time to time, computing equipment has to be moved to a new location; the smaller the size of the computing equipment, the lower the transportation costs of moving from one location to the other. (d) Communication delays do matter; the smaller the size of the computing equipment, the shorter the communication delays.

Electricity costs of the computing production do matter, for the following reasons: (a) Except for the salaries of system administrators and application programmers, some studies reveal that 50 % of operational costs is due to the electricity costs. (b) Lower electricity consumption means less expenses for cooling. (c) Lower electricity consumption related to heating means less wear and tear of the infrastructure around the production computers. (d) Less electricity means less hazards for the natural environment around.

Based on the previously stated facts about the DataFlow supercomputing, one can conclude that the DataFlow programming will not only result in a faster code but also in lower operational costs, both due to smaller equipment size and smaller energy consumption.

1.14   Computer Size Matters (1U, 2U, 4U)

The most appropriate way to compare computer systems is to compare the speed of the computing structure that can fit into a 1U box, which is the unit size for modules of a rack-mount supercomputer system. In that case, size and power are kept identical for all systems to be compared, and consequently, the speed comparison is the most fair. It is exactly such a comparison approach that favors the DataFlow system. So, the question is: What is 1U? Also, another question is: What are 2U and

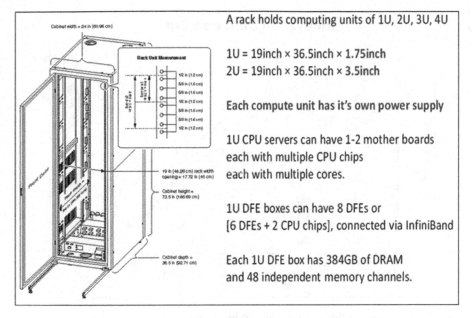

Fig. 1.21 Facts about 1U, 2U, and 4U (Courtesy of Maxeler Technologies)

4U, as derivatives of 1U? Figure 1.21 illustrates what can be contained in a 1U, from the point of view of Maxeler Technologies, while recalling the related terminology about racks.

1.15 Bottom Line

When writing a line of code, the question is how much operational costs one creates with that line. From the aforementioned, the created operational cost is much smaller for DataFlow machines than for control flow machines. Consequently, one can say that DataFlow programming is *OPEX-aware programming*.

Bearing in mind all of the above, it is wise to be an early adopter and to join the DataFlow user family promptly. Figure 1.22 shows the innovation diffusion curve. The DataFlow technology is now at the tipping point, meaning that early adopters maximize the benefits of the adopted technology.

The tipping point coincides with the introduction of the OpenSPL – a creation that makes the DataFlow concept easily usable. Till that point, the community was aware of the speedups in hardware and was suspicious about the effectiveness in programming. It is the introduction of the OpenSPL that created the full trust into the new technology: great speedup and easy programming. Figure 1.23 recalls the major sponsors of the OpenSPL standard as announced in the related website.

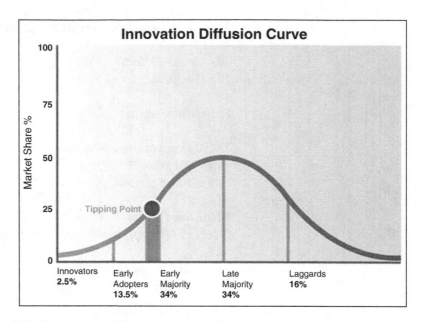

Fig. 1.22 The innovation diffusion curve and the tipping point

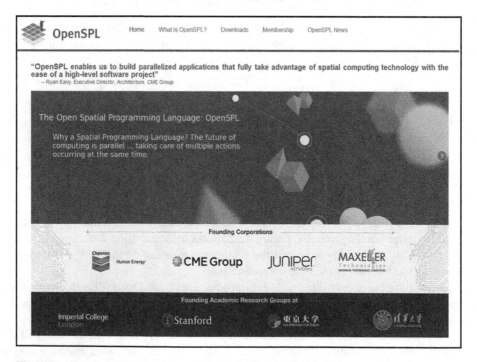

Fig. 1.23 The OpenSPL Consortium (Courtesy of Maxeler Technologies)

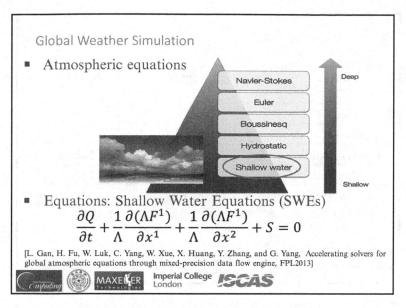

Fig. 1.24   Results from a Shallow Water Weather Forecast model [Gan2013]

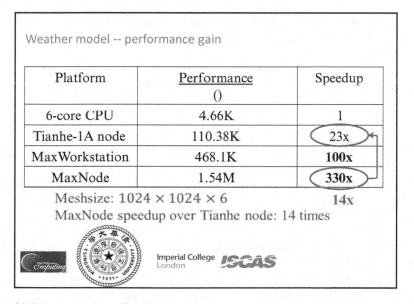

Fig. 1.25   More results from [Gan2013]

Figures 1.24, 1.25, and 1.26 summarize the superior performance of a DataFlow machine over the control flow machine on the top of Green500, of course on the level of 1U, which represents the fairest way of comparison, as already indicated. The example from these three figures points to one important (previously mentioned)

Weather model -- power efficiency

Platform	Efficiency ( )	Speedup
6-core CPU	20.71	1
Tianhe-1A node	306.6	14.8x
MaxWorkstation	2.52K	**121.6x**
MaxNode	3K	**144.9x**

Meshsize: $1024 \times 1024 \times 6$ 9 x

MaxNode is 9 times more power efficient

Imperial College London ISCAS

Fig. 1.26 Still more self-explanatory results from [Gan2013]

advantage of the DataFlow concept: the floating-point precision was reduced, from 32 bits to 15 bits, at no jeopardy as far as the quality of final results was concerned [Gan2013].

One can find more on the potentials of the DataFlow architecture in [Flynn2013] and on the Web [e.g., Google: Maxeler Technologies].

1.16 OpenSPL

Basic notions of OpenSPL are given in Fig. 1.27. The major issue is that the DataFlow paradigm requires programmers to think in space, rather than in time!

Basic motivations for programming in space are enlisted in Fig. 1.28. The major issue is that the growth of on-chip transistor count cannot be fully exploited due to the limited memory bandwidth. Consequently, as elaborated in Fig. 1.29, the solution is in the DataFlow approach.

The major goals of OpenSPL are underlined in Fig. 1.30, while the semantic structure of OpenSPL is described in Fig. 1.31. Three OpenSPL examples are given in Figs. 1.32, 1.33, and 1.34. All the examples are built to be self-explanatory.

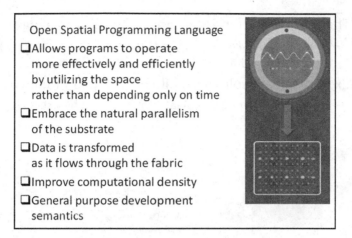

Open Spatial Programming Language

❑Allows programs to operate
   more effectively and efficiently
   by utilizing the space
   rather than depending only on time

❑Embrace the natural parallelism
   of the substrate

❑Data is transformed
   as it flows through the fabric

❑Improve computational density

❑General purpose development
   semantics

Fig. 1.27 Capabilities of OpenSPL (Courtesy of Maxeler Technologies)

❑ Core clock frequencies evened out in the few GHz range

❑ Energy / Power consumption of modern HPC systems
   became huge economic burden not to be ignored any longer

❑ Specialization has proven its power efficiency potentials

❑ The requirements for annual performance improvements keep growing steadily

❑ SoCs are now exploiting also the third dimension (3D-int)

❑ However, the majority of programmers build upon the legacy,
   1D linear view and sequential execution

❑ Many clever proposals but no good solution to date
   (e.g., Cilk, Sequoia, OmpSs and OpenCL)

Fig. 1.28 Basic motivations for programming in space

1.17   WebIDE

A major recent development oriented to customer satisfaction, in the case of
Maxeler, is the WebIDE integrated development system. Snapshots of its usage,
showing the elegancy of the Web-based approach for program development, are
given in Figs. 1.35, 1.36, 1.37, 1.38, and 1.39. One can select to run an existing
application or to create a new one, using a number of ready-to-use examples. One
can build kernels using these examples, and one can run the application remotely
on a real Maxeler machine, using a cloud. More information can be found at the
website http://webide.maxeler.com.

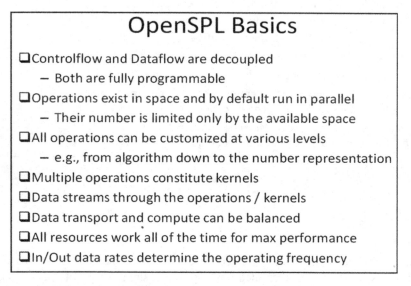

Fig. 1.29 Technology-related motivations for programming in space (Courtesy of Maxeler Technologies)

> # OpenSPL Basics
>
> ❑ Controlflow and Dataflow are decoupled
> - Both are fully programmable
> ❑ Operations exist in space and by default run in parallel
> - Their number is limited only by the available space
> ❑ All operations can be customized at various levels
> - e.g., from algorithm down to the number representation
> ❑ Multiple operations constitute kernels
> ❑ Data streams through the operations / kernels
> ❑ Data transport and compute can be balanced
> ❑ All resources work all of the time for max performance
> ❑ In/Out data rates determine the operating frequency

Fig. 1.30 The basics of OpenSPL (Courtesy of Maxeler Technologies)

1.18   Instead of the Conclusion

If there is a single pearl of wisdom to get out of this book, then it is the link between the DataFlow supercomputing and the theoretical work of the Nobel Laureate Richard Feynman [Feynman96]: arithmetic and logic could be done with a zero or almost zero energy; communications cannot. The control flow approach is based on

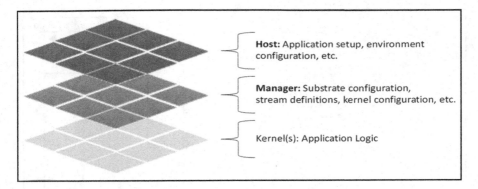

Fig. 1.31 The structure of OpenSPL (Courtesy of Maxeler Technologies)

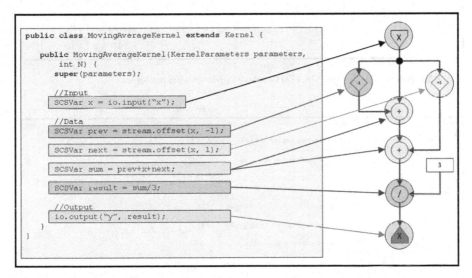

Fig. 1.32 An example of OpenSPL (1); remake of a previous figure, with features of OpenSPL underlined (Courtesy of Maxeler Technologies)

the von Neumann paradigm, which implies lots of communications, as indicated in Fig. 1.40 (left side). One first fetches an instruction, which implies communications to the program memory. Then one fetches data in two different communication-based transactions, from two different memory locations. Finally, one stores the result into the data memory, which means one more communication activity.

In the case of DataFlow, as indicated in Fig. 1.40 (right side), ideally there are no communications: DataFlow from one arithmetic or logic unit to the other. Of course, this ideal case is obtainable only if the compiler is smart enough to generate a planar graph and only the zero length communication paths between two neighboring arithmetic or logic units.

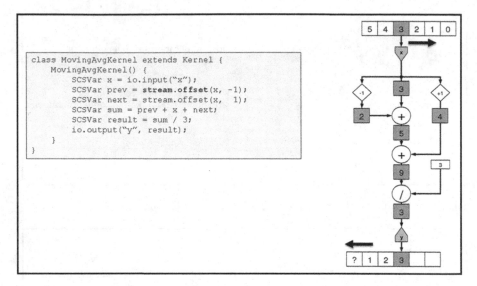

```
class MovingAvgKernel extends Kernel {
    MovingAvgKernel() {
        SCSVar x = io.input("x");
        SCSVar prev = stream.offset(x, -1);
        SCSVar next = stream.offset(x,  1);
        SCSVar sum = prev + x + next;
        SCSVar result = sum / 3;
        io.output("y", result);
    }
}
```

Fig. 1.33 An example of OpenSPL (2) (Courtesy of Maxeler Technologies)

```
// Based on: "A New Formula for Computing Implied Volatility" by Steven Li
SCSVar impliedVol(SCSVar optionPrice,
                  SCSVar futurePrice,
                  SCSVar strikePrice,
                  SCSVar timeToExpiration,
                  SCSVar interestRate) {

    SCSVar discountFactor = exp(interestRate*timeToExpiration);

    optionPrice = optionPrice * discountFactor;

    SCSVar sqrtT = sqrt(timeToExpiration);

    SCSVar KmS = strikePrice - futurePrice;
    SCSVar SpK = futurePrice + strikePrice;

    SCSVar alpha = (sqrt(2.0*Math.PI) / SpK) * (optionPrice + optionPrice + KmS);

    SCSVar tempB = max(0, alpha*alpha - 4.0*KmS*KmS/(futurePrice*SpK));

    return 0.5*(alpha + sqrt(tempB)) / sqrtT;
}
```

Running time: ~700ns

Fig. 1.34 An example of OpenSPL (3) (Courtesy of Maxeler Technologies)

Fig. 1.35 The welcome page, WebIDE (Courtesy of Maxeler Technologies)

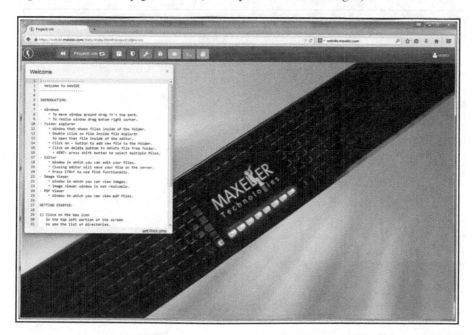

Fig. 1.36 The getting started, WebIDE (Courtesy of Maxeler Technologies)

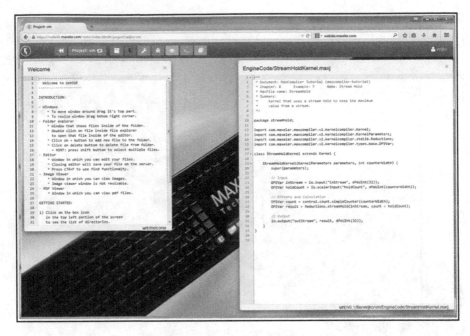

Fig. 1.37 The kernel code, WebIDE (Courtesy of Maxeler Technologies)

Fig. 1.38 The manager code, WebIDE (Courtesy of Maxeler Technologies)

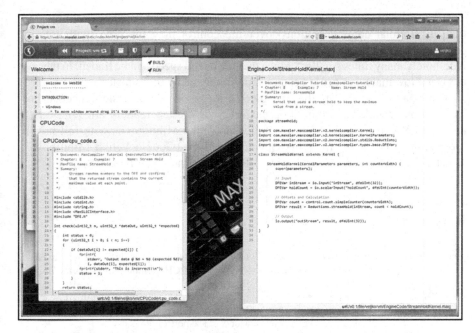

Fig. 1.39   The build and run, WebIDE (Courtesy of Maxeler Technologies)

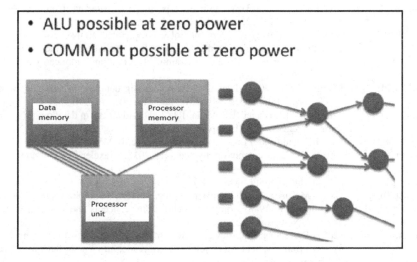

Fig. 1.40   Essence of the Feynman-based explanation of DataFlow potentials

Consequently, in the ideal case, the power consumption of the DataFlow machine (due to minimal communications) over the control flow machine (with lots of communications) becomes much smaller (in the ideal case, the power could be indefinitely many times smaller). Since the power and speed can be traded, in theory, the DataFlow machines could become indefinitely many times faster!

This analysis concludes the first part of the book, about basic concepts. For specific and detailed programming examples, interested readers are referred to the website with courses taught by Professor Veljko Milutinović, with full-blown programming examples from his OpenCourseWare (Google Veljko Milutinović; select TEACHING; next select VLSI). URL = http://home.etf.rs/~vm/

References

[Dennis1974]   Dennis JB, Misunas DP (1974) A preliminary architecture for a basic data-flow processor. Newsl ACM SIGARCH Comput Archit News Homep 3(4):126–132

[DOE2014]   US Department of Energy (2014) Advanced scientific computing research - X-Stack portfolio, April [Online]. Available: http://science.energy.gov/ascr/research/computer-science/ascr-x-stack-portfolio/

[Dongarra94]   (2014) *Top500* [Online]. Available: http://en.wikipedia.org/wiki/TOP500

[Feynman96]   Feynman PF (1996) Feynman lectures on computation. Addison-Wesley Publishing Company Inc., Boston

[Flynn2013]   Flynn M et al (2013) Moving from petaflops to petadata. Commun ACM 56(5):39–42. ACM, New York, NY

[Gan2013]   Gan L et al (2013) Accelerating solvers for global atmospheric equations through mixed-precision dataflow engine. In: Proceedings of the 23rd international conference on Field Programmable Logic and applications (FPL), Porto, Portugal, pp 1–6

[Giorgi2014]   Giorgi R et al (2014) TERAFLUX: harnessing dataflow in next generation teradevices. Microprocess Microsyst 38(8, Part B):976–990

[Johnston2004]   Johnston WM, Hanna JRP, Millar RJ (2004) Advances in dataflow programming languages. ACM Comput Surv 36(1):1–34

[Linux2000]   Siever E et al (2009) Linux in a nutshell. O'Reilly Media, Sebastopol

[Maxeler2012]   (2012) Exascale computing by the year 2018. Maxeler Technologies Ltd, London

[Maxeler2014]   (2014) The OpenSPL. Maxeler Technologies Ltd, London

[Maxeler2015]   (2015) Multiscale dataflow programming. Maxeler Technologies Ltd, London

[Milutinovic88]   Milutinovic V (1988) Computer architecture: concepts and systems. North-Holland, New York

[Milutinovic2014]   (2014) BALCON: the gateway to ICT monitoring & control research in the Western Balkans, September [Online]. Available: http://www.balcon-project.eu/mainpage

[Moskowitz2014]   Moskowitz H (2007) Selling blue elephant. Pearson Education Inc. Publishing as Prentice Hall, Upper Saddle River

[Nemeth2008]   Nemeth T et al (2008) An implementation of the acoustic wave equation on FPGAs. In: Proceedings of the 78th Society of Exploration Geophysicists (SEG) meeting, Las Vegas, November 2008, pp 2874–2878

[SEG2008]   (September, 2014) [Online]. Available: http://home.etf.rs/~vm/os/mips/MaxelerBelgradeTalkAugust10.pdf

[STFC2014]   (2014) STFC Daresbury Laboratory first to install maximum performance
                    computer (MPC), February 25 2014 [Online]. Available: http://www.maxeler.
                    com/stfc-dataflow-supercomputer
[Stojanovic2013]   Stojanovic S, Bojic D, Milutinovic V (2013) Solving Gross Pitaevskii equa-
                    tion using dataflow paradigm. IPSI Trans Internet Res Belgrade 9(2):19–22,
                    Serbia
[Stojanovic2015]   Stojanovic S, Milutinovic V (2015) A survey of dataflow architectures. Adv
                    Comput Elsevier 96:1–45
[WANG2010]   Wang YH (2010) Multichannel matching pursuit for seismic trace decomposi-
                    tion, In Proceedings of the 72nd EAGE conference & exhibition incorporating
                    SPE EUROPEC 2010, Barcelona, Spain, June 2010

Chapter 2
Selected Case Studies

2.1 Classification of Selected Examples

The classification methods that are used in this analysis were developed under the impact of the European Union FP7 environment dealing with next-generation Big Data management solutions.

Application and Sub-application Criteria

The classification solely depends on criteria used. In this chapter the authors suggest two application criteria and one sub-application criterion to present two classifications and one sub-classification.

1. The first and the main criterion is *what science or industry the application is targeted for*. According to this criterion, there are three groups of applications:

   (a) *Exascale Fundamental Drivers* – targeting Formal Sciences (Decision Theory, Logic, Mathematics, and Statistics), Finances, Transportation, and algorithms pertaining to FPGA improvements
   (b) *Exascale Science and Technology Drivers* – targeting Natural Sciences (Life Sciences, Physical Sciences, and Earth Sciences), Social Sciences, and Medicine
   (c) *Exascale Engineering and Innovation Drivers* – targeting Engineering and different industries

   This will be the main classification used in the paper. It consists of the group name (as a tree's branch) and the exact name of the science or industry (as the branch's leaf).
   The other two, classification and subclassification, will be just mentioned.

© Springer International Publishing Switzerland 2015
V. Milutinović et al., *Guide to DataFlow Supercomputing*, Computer
Communications and Networks, DOI 10.1007/978-3-319-16229-4_2

2. The second criterion is what type of task the DataFlow algorithms in the application perform. The algorithm, according to its role, is placed in one of the following five groups of tools:

   (a) *Optimization toolbox*
   (b) *Complex networks analysis toolbox*
   (c) *Image, video, text processing, and analysis toolbox*
   (d) *Numerical analysis, modeling, and simulation toolbox*
   (e) *Machine learning and data mining toolbox*

3. The sub-application criterion is used just for the applications that have originally been made for other computer architectures (CPU or GPP). According to what type of adaptation was performed during the application transfer procedure, there are three groups: (i) a particular single algorithm or (ii) a set of algorithms (framework) or (iii) a set of related algorithms composed as a complete stand-alone solution. Additionally, according to what each algorithm targets, there are two groups: (a) a specific application or (b) a targeted set of applications.

2.1.1   Examples for Each Leaf Within the Main Classification

Exascale Fundamental Drivers

Examples of the Leaf #1 Finance

1. A Mixed Precision Monte Carlo Methodology for Reconfigurable Accelerator Systems [Chow2012]
2. Finding the Right Level of Abstraction for Minimizing Operational Expenditure [Mencer2011]
3. Accelerating Reconfigurable Financial Computing [Tse2012]
4. Accelerating the Computation of Portfolios of Tranched Credit Derivatives [Weston2010]
5. Multi-level Customization Framework for Curve Based Monte Carlo Financial Simulations [Jin2012]
6. Rapid Computation of Value and Risk for Derivatives Portfolios [Weston2011]

Examples of the Leaf #2 Mathematics

1. A Fully-Pipelined Expectation-Maximization Engine for Gaussian Mixture Models [Guo2012]
2. Enhancing Performance of Tall-Skinny QR Factorization Using FPGAs [Rafique2012]
3. Heterogeneous Reconfigurable System for Adaptive Particle Filters in Real-Time Applications [Chau2013]
4. Optimizing Performance of Quadrature Methods with Reduced Precision [Tse2012/2]

5. Customizable Architectures for the Set Covering Problem [Guo2013]
6. A Fully Pipelined Probability Density Function Engine for Gaussian Copula Model [Ruan2014]

Exascale Science and Technology Drivers

Examples of the Leaf #1 Biology

1. Hardware Acceleration of Genetic Sequence Alignment [Arram2013]
2. A Large-Scale Spiking Neural Network Accelerator for FPGA Systems [Cheung2012]

Examples of the Leaf #2 Geophysics

1. Maximum Performance Computing with DataFlow Engines [Pell2012]
2. An Implementation of the Acoustic Wave Equation on FPGAs [Nemeth2008]
3. Finding Speedup in Parallel Processors [Flynn2008]
4. Anisotropic Reverse-Time Migration Using Co-Processors [Liu2009]

Examples of the Leaf #3 Meteorology

1. Acceleration of a Meteorological Limited Area Model with DataFlow Engines [Oriato2012]

Exascale Engineering and Innovation Drivers

Examples of the Leaf #1 Oil and Gas Industry

1. Surviving the End Of Frequency Scaling with Reconfigurable DataFlow Computing [Pell2011]
2. Beyond Traditional Microprocessors for Geoscience High-Performance Computing Applications [Lindtjorn2011]
3. Acceleration of Anisotropic Phase Shift Plus Interpolation with DataFlow Engines [Tomas2012]
4. Accelerating Large-Scale HPC Applications Using FPGAs [Dimond2011]
5. Accelerating 3D Convolution Using Streaming Architectures On FPGAs [Fu2009]
6. Finite-Difference Wave Propagation Modeling on Special-Purpose DataFlow Machines [Pell2013]

2.2 Presentation of Examples

In this section, the authors present examples of the transferred applications that showed reasonable speedups and are found in available open literature. For each classification leaf, one, the most representative example, is described in more detail, while for the rest of them, only a short description is given together with acquired results (speedup, power consumption reduction, and other improvements).

2.2.1   Examples from Exascale Fundamental Drivers Leaf #1: Finance

2.2.1.1   Example 1: A Mixed Precision Monte Carlo Methodology for Reconfigurable Accelerator Systems [Chow2012]

Classification #2 – Numerical analysis, modeling, and simulation toolbox
Subclassification – A particular single algorithm for a targeted set of applications
Applications/algorithms – Monte Carlo simulation

In this article, Gary C. T. Chow et al. introduce a novel method of mixed precision applicable to any Monte Carlo (MC) simulation. The MC simulations can be well used in field-programmable gate arrays. They have parallel nature that can be well executed in FPGAs, and there are, also, the needed cost-effective Maxeler DFE random number generators. The authors used data paths with reduced precision, and for correcting any resulting errors, they used auxiliary sampling. To determine the optimal reduced precision and optimal resource allocation among the MC data paths and correction data paths, optimization based on mixed integer geometric programming was implemented. Reduced-precision data paths usually have higher clock frequencies, consume fewer resources, and offer higher degree of parallelism, for a given amount of resources compared with full-precision data paths.

The results were evaluated for three financial use cases: Asian option pricing, fixed strike lookback call option under the GARCH model, and collateralized mortgage obligation, and in multidimensional integral evaluation that is used in many other areas.

Essence
The major contributions of this paper are:

– Techniques for partitioning workloads of different precisions for auxiliary sampling to a reconfigurable accelerator system consisting of FPGA(s) and GPP(s)
– Mixed Integer Geometric Programming for finding optimal precision for FPGA's data paths and optimal resource allocation and an optimization method for the execution time of a Monte Carlo simulation on a reconfigurable accelerator system based on an analytical model
– A novel mixed precision methodology for correcting finite precision errors using auxiliary sampling and an error analysis that indicates finite precision errors from sampling errors in reduced precision Monte Carlo simulations

Infrastructure
The system architecture is presented in Fig. 2.1

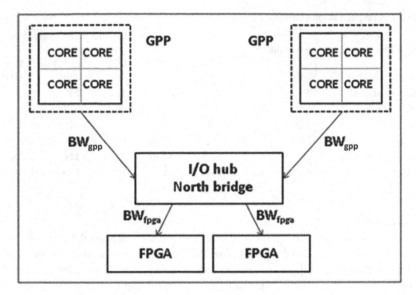

Fig. 2.1 System architecture of the reconfigurable accelerator system in the analytical model [Chow2012]

Relevance and Details

The proposed mixed precision methodology provides several advantages over previous FPGA designs.

– Approximated mean finite precision error μ *fin* is used in adjusting the final result. Instead of passively finding the error bound, this novel approach enables obtaining a more accurate result from the reduced precision result.
– A very accurate result can be achieved by increasing the number of sample points because there are only sampling errors in the output. The output accuracy is no longer bound by the reduced precision.
– Since this methodology is totally independent of the function and no accuracy analysis is required for the relative error, the methodology is applicable to any Monte Carlo simulation.

Application and Results

The achieved results are very remarkable (Table 2.1). Mixed precision FPGA reconfigurable accelerator system is up to 4.6 times faster than state-of-the-art GPU, 7.1 times faster than an FPGA using double precision, and 163 times faster than optimized software on a quad-core GPP. As far as energy efficiency is concerned, it is up to 5.5 times more energy efficient than a GPU and 170 times more energy efficient than a quad-core GPP.

Table 2.1 Comparison with GPP, GPU, and plain FPGA [Chow2012]

Type of MC simulation:	Asian option				Collateralized mortgage obligation		
System configuration	GPP only	GPU	FPGA only	FPGA + GPP	GPP only	FPGA only	FPGA + GPP
Execution time (s)	29	3	4.7	0.56	117	2.8	0.72
Energy (kJ)	5.3	0.71	0.4	0.13	20,4	0,26	0,12
Power (W)	183	236	85	192	175	94	171
Precision	Double	Double	Double	Mixed	Double	Double	Mixed
Normalized speedup	**1x**	**9.7x**	**6.2x**	**44.6x**	**1x**	**42x**	**163x**
Normalized energy	**40.7x**	**5.5x**	**3.1x**	**1x**	**170**	**2.2x**	**1x**

Characteristics

This work shows that very good improvements can be achieved by using lower precision. Of course, in order to be sure that results have satisfactory precision, one must have numerically exact references of previously done calculations. The authors had the reference results available, so having data to compare with; they could perform all the tests and get precise results.

Trends

Further analysis with changing data precision on FPGA was continued and has been tried by many authors. General-purpose computer architectures do not offer data precision changing, so experimenting with it on DataFlow computers was to be expected.

2.2.1.2 Example 2: Finding the Right Level of Abstraction for Minimizing Operational Expenditure [Mencer2011]

Classification #2 – Numerical analysis, modeling, and simulation toolbox
Subclassification – A set of algorithms (framework) for a targeted set of applications
Applications/algorithms – Total cost of ownership (TCO) of a financial computing
 operation, Monte Carlo case study

In this article, Oskar Mencer et al. were examining how modern programming language abstractions impact total cost of ownership (TCO) of a financial computing operation. The study was mostly done using static and dynamic analysis of financial software example based on the loopflow graph (LFG) concept and the custom dynamic hotspot tool called MaxSpot.

In the real life, companies' corporate structures are usually not very keen on releasing correct TCO calculations or figures. The authors suggested the following list of components:

(a) Cost of software development and testing; (b) Capital expenditure (CAPEX): cost of computing equipment, cost of air-conditioning, cost of computer storage, cost of communication equipment, and cost of data center real estate; (c) Operational expenditure (OPEX): cost of datacenter staff, cost of electricity for computers and air-conditioning, and cost of real estate operations; (d) Indirect costs: cost of failures (unpredicted and scheduled downtime and nonavailability causing business decreases), cost of time while determining whether there was a human or technical failure, and cost of reliability (testing and verification of computing, backups, redundancy).

It is in all businesses difficult to calculate what the real costs of not getting the needed data on time are. It's even more so in the financial industry (the fast changes on the financial instrument markets, the interest rates, etc.).

The authors took a simple Monte Carlo application in C and reprogrammed it in C++ to enable raising the level of abstraction (adding payoff calculations and random number generators) making expanding of the code to be done with little programming effort.

The most valuable result of these analyses is the conclusion that *provided the required throughput of an application is high enough, the operational expenditure decreases through minimizing run time and not through minimizing programming effort.*

2.2.1.3 Example 3: Accelerating Reconfigurable Financial Computing [Tse2012]

Classification #2 – Numerical analysis, modeling, and simulation toolbox
Subclassification – A set of related algorithms composed as a complete stand-alone solution, for a targeted set of applications
Applications/algorithms – Derivative pricing using both Monte Carlo and quadrature methods

This PhD thesis proposes novel approaches for designing, optimization, and management of reconfigurable compute accelerators for financial computing:

1. Proposed novel reconfigurable designs for derivative pricing using both Monte Carlo and quadrature methods. Such designs involved exploring different techniques such as multidimensional analysis for quadrature methods and control variate optimization for Monte Carlo. By using the field-programmable gate array (FPGA) designs, significant speedups and energy savings over both CPU and GPU designs were achieved.

2. Proposed novel framework for distributing computing tasks on multi-accelerator heterogeneous clusters. In this framework, different computational devices including FPGAs, GPUs, and CPUs were used. They worked collaboratively on the same financial problem based on a dynamic scheduling policy. The author investigated trade-offs in speed and in energy consumption of different accelerator allocations.
3. Proposed new reduced precision methodology for optimizing quadrature designs and a mixed precision methodology for optimizing Monte Carlo designs. The optimized throughput of reconfigurable designs was achieved by using data paths with minimized precision while carefully obtaining the same accuracy of the results as in the original designs.

This work showed trade-offs between the contribution of each computation in increasing the accuracy of the final result and the number of computations. When mixed precision methodologies were used, the speedups of 2.9 to 7.1 times over the double-precision FPGA designs and 44 to 106 times over the quad-core CPU designs were achieved. Using the mixed precision methodology, a Virtex-6 ST475X FPGA and an i7-870 CPU were able to outperform 448-core GPU by 4.6 times while using 5.5 times less energy.

2.2.1.4   Example 4: Accelerating the Computation of Portfolios of Tranched Credit Derivatives [Weston2010]

Classification #2 – Optimization toolbox
Subclassification – A set of related algorithms composed as a complete stand-alone solution, for a targeted set of applications
Applications/algorithms – Standard base correlation methodology, with a Gaussian copula for default correlation and a stochastic recovery process

Recent big growth in trading and complexity of credit derivative instruments has reasonably increased the need for more computationally demanding mathematical models. Since today's usual way to provide for those increases is big-sizing the computer centers, this has led to massive growth in data center compute capacity, power, and cooling requirements. The paper reports the results of a joint project between J.P. Morgan and Maxeler Technologies on improving the price/performance for calculating the value and risk of a large complex credit derivatives portfolio.

The results showed that valuing tranches of collateralized default obligations (CDOs) on Maxeler accelerated systems was over 30 times faster per Watt and per cubic foot than solutions using standard multi-core Intel Xeon processors. Also reports on some preliminary results of further work that extends the approach to classes of interest rate derivatives were given.

2.2.1.5 Example 5: Multi-level Customization Framework for Curve Based Monte Carlo Financial Simulations [Jin2012]

Classification #2 – Numerical analysis, modeling, and simulation toolbox
Subclassification – A particular single algorithm for a targeted set of applications
Applications/algorithms – Monte Carlo framework for the automated generation of
   highly efficient curve-based payoff evaluation accelerator

One of the main challenges when accelerating financial applications using reconfigurable hardware is the management of design complexity. This paper proposed a multi-level customization framework for automatic generation of complex yet highly efficient curve-based financial Monte Carlo simulators on reconfigurable hardware. By identifying multiple levels of functional specializations and the optimal data format for the Monte Carlo simulation, different levels of programmability were allowed in the framework to retain good performance and support multiple applications.

The results showed that Virtex-6 SX475T FPGAs generated by this framework were about 40 times faster than single-core software implementations on an i7-870 quad-core CPU at 2.93 GHz; they were also over 10 times faster and 20 times more energy efficient than 4-core implementations on the same i7-870 quad-core CPU and were over three times more energy efficient and 36 % faster than a highly optimized implementation on an NVIDIA Tesla C2070 GPU at 1.15 GHz.

What is interesting is the fact that this framework is platform independent and can be extended to support CPU and GPU applications.

2.2.1.6 Example 6: Rapid Computation of Value and Risk for Derivatives Portfolios [Weston2011]

Classification #2 – Machine learning and data mining toolbox
Subclassification – A set of related algorithms composed as a complete stand-alone
   solution for a specific application
Applications/algorithms – Valuation to risk measurement and multivariate Monte
   Carlo derivative pricing model

This paper reports on results of another project between J.P. Morgan in London and Maxeler Technologies on accelerating derivatives computations. Compared to the previously mentioned project, in this one, the work was extended in two ways: by applying the same techniques, first, to accelerate the computation of portfolio-level risk for credit derivatives and, second, to different asset classes using a different type of mathematical model. Also the implications for risk were explored.

The paper quotes an interesting information that in 2005, the world's estimated 27 million servers consumed around 0.5 % of all electricity produced on the planet, a figure that is closer to 1 % when the energy for associated cooling and auxiliary equipment (e.g., backup power, power conditioning, power distribution, air handling, lighting, and chillers) is included.

A MaxNode-1821 compared to an eight-core Xeon E5430 server showed 31 times speedup in full precision and 37 times in reduced precision while the power usage per node was decreased by 3 %.

2.2.2 Examples from Exascale Fundamental Drivers Leaf #2: Mathematics

2.2.2.1 Example 1: A Fully-Pipelined Expectation-Maximization Engine for Gaussian Mixture Models [Guo2012]

Classification #2 – Machine learning and data mining toolbox
Subclassification – A particular single algorithm for a targeted set of applications
Applications/algorithms – Gaussian mixture models for probability density model-
  ing and soft clustering

In this text, Ce Guo et al. describe Gaussian mixture models (GMMs), a powerful tool for soft clustering and probability density modeling, implemented using Maxeler DFE. In many such applications, it is necessary to estimate parameters of a GMM from data before working with it. One way to handle this task is to use computationally demanding expectation-maximization algorithm for Gaussian mixture models (EM-GMM). In order to fully benefit from the FPGA pipelined hardware architecture, the authors proposed a pipeline-friendly EM-GMM algorithm. They also used a Gaussian probability density function evaluation unit (working with fixed-point arithmetic) to further improve the performance.

Time spent on each EM-GMM algorithm run directly depends on the data size. Due to a fast growth of data size (big data), the algorithm has became very computationally demanding. At the same time, it was desirable for the algorithm to be executed within a short period of time, especially in real-time applications.

Essence
The major contributions of this paper are:

– Making algorithmic transformations in the structure of the work flow of the original EM-GMM algorithm in order to enable pipelining of different computation
  stages, creating a pipeline-friendly EM-GMM algorithm
– Suggesting customized design of the Gaussian probability density evaluation unit
  that minimizes the hardware cost while obtaining satisfactory accuracy
– Solving precision problem in Gaussian PDF evaluation with bit shifting and
  successfully deploying fixed-point arithmetic throughout the system

Infrastructure
In the performance experiments, the systems were compared with a GPU implementation described in [Kumar2009]. Since the authors did not have experimental results of that system on the same datasets they used, they took the best performance records in [Kumar2009] with similar data sizes. That means that there is a slight possibility that the performance estimation could stray and that the GPU comparison results were provided in the paper only for reference.

The two CPU implementations were done on Intel Core i3 CPU (running at 2.93 GHz) and 4 GB DDR3 memories. The FPGA implementation was deployed on a Maxeler MAX3 acceleration card with a Xilinx Virtex-6 FPGA running at 150 MHz and with 48GB DDR3 onboard memory.

Relevance and Details
Gaussian Mixture Models (GMMs) are powerful tools for probability density modeling and soft clustering. Some relevant usages quoted in the paper are (1) an image segmentation system that identifies tissues in magnetic resonance (MR) images of the brain where GMMs are employed to capture the spatial layout information of brain tissues, (2) a speaker verification system where GMMs are used to model the characters of a speaker's voice, and (3) a computer vision system to cut the background from video streams where GMMs are used to judge whether a pixel belongs to the background in a probabilistic manner.

The fundamental difference between the pipeline-friendly EM-GMM and the original EM-GMM is that the former requires data to be streamed into the algorithm only once, while the latter requires it three times. Other important differences are:

- The maximization step and the expectation step become overlapped in the pipeline-friendly algorithm. In the original algorithm, first all the data instances are processed in the expectation step and only then the maximization step starts. In the pipeline-friendly algorithm, the dataset is handled in a per-instance manner. Statistical information of the new parameter set is updated when the data instance arrives.
- The original algorithm stores all the responsibility values in the expectation step. In the pipeline-friendly algorithm, it is not necessary to do that because the algorithm computes the responsibility values for a newly arrived data instance and right away updates the statistical information of the new parameter set. The corresponding responsibility values can be discarded safely as soon as the statistical information about a data instance is collected.

Application and Results
The algorithm performance is measured by the number of data instances processed in every second. A data instance is considered to be processed "in one iteration" after all computations related to it are completed in that iteration. Experimental results on

Table 2.2 Performance
results (instances per second)
[Guo2012]

Data	CPU1	CPU2	GPU	FPGA	SU$_{C1}$	SU$_G$
1	1.723	2.040	3.081	1.498	86x	5x
2	8.565	1.024	1.541	1.492	174x	9x
3	5.689	6.671	1.027	1.402	262x	15x
4	8.714	1.023	1.641	1.487	171x	9x
5	4.310	5.079	7.704	1.487	347x	15x
6	2.883	3.402	5.136	1.487	517x	28x

performance are presented in Table 2.2. The last two columns in the table are the
speedup values of FPGA-based solution over the CPU-based solution (the original
EM-GMM) and the GPU-based solution, respectively.

The results showed that the FPGA-based solution generated rather accurate
results and achieved a maximum of 517 times speedup over a CPU-based solution
and 28 times speedup over a GPU-based solution.

Characteristics
To reach such high accelerations, it is necessary to use fixed-point arithmetic that
saves hardware resources on the FPGA platform. This enables deployment of up to
36 Gaussian PDF evaluation units in the pipeline, excellently using available FPGA
resources. However, it is not feasible to perform similar optimization on CPUs and
GPUs. With richer logical resources on the FPGA platform, it would be possible
to deploy even larger number of Gaussian PDF evaluation units enabling more
complicated data to be processed by the system. The corresponding acceleration
would be even more significant.

Trends
This example shows possible advantages of FPGA architecture in algorithms that
work satisfiably with fixed-point arithmetic.

2.2.2.2   Example 2: Enhancing Performance of Tall-Skinny QR Factorization Using FPGAs [Rafique2012]

Classification #2 – Optimization toolbox
Subclassification – A particular single algorithm for a targeted set of applications
Applications/algorithms – Tall-skinny QR factorization

Tall-skinny QR factorization (TSQR) is one of the communication-avoiding
linear algebra algorithms with low communication latency and high memory
bandwidth requirements. As such, it is an excellent candidate for acceleration using
FPGAs. TSQR parallelizes QR factorization of tall-skinny matrices in a divide-
and-conquer fashion by decomposing them into sub-matrices, performing local QR

factorizations, and then merging the intermediate results. GPUs seem to be the hardware of choice (performance-wise) for this algorithm since it is a dense linear algebra problem. However, memory bandwidth in local QR factorizations and global communication latency in the merge stage limit the performance of GPUs.

In this paper, the shape of the matrix was exploited on FPGA-based custom architecture, which avoided these bottlenecks by using high-bandwidth on-chip memories for local QR factorizations and by performing the merge stage entirely on-chip to reduce communication latency.

The result that the authors achieved was a peak double-precision floating-point performance of 129 GFlops on Virtex-6 SX475T. A quantitative comparison of the proposed design with recent QR factorization on FPGAs and GPU showed speedup of up to 7.7 and 12.7 times, respectively. Additionally, even higher performance over optimized linear algebra libraries like Intel MKL for multi-cores, CULA for GPUs, and MAGMA for hybrid systems was achieved.

2.2.2.3 Example 3: Heterogeneous Reconfigurable System for Adaptive Particle Filters in Real-Time Applications [Chau2013]

Classification #2 – Optimization toolbox
Subclassification – A particular single algorithm for a targeted set of applications
Applications/algorithms – Particle filter statistical method for dealing with dynamic systems having nonlinear and non-Gaussian properties

This paper presents a heterogeneous reconfigurable system for real-time applications applying particle filters. The system consists of a multi-threaded CPU and an FPGA. The authors proposed a method to adapt the number of particles dynamically and to use run-time reconfigurability of the FPGA to reduce power and energy consumption. An application was developed that involves simultaneous mobile robot localization and people tracking. The results showed that the proposed adaptive particle filter could reduce up to 99 % of computation time.

Using run-time reconfiguration, a reduction of 34 % in idle power and 26–34 % of system energy was achieved. The proposed system was up to 7.39 times faster and 3.65 times more energy efficient than Intel Xeon X5650 CPU with 12 threads and 1.3 times faster and 2.13 times more energy efficient than NVIDIA Tesla C2070 GPU.

2.2.2.4 Example 4: Optimizing Performance of Quadrature Methods with Reduced Precision Tse2012/2]

Classification #2 – Optimization toolbox
Subclassification – A particular single algorithm for a targeted set of applications
Applications/algorithms – Quadrature methods

This paper presents a generic precision optimization methodology for quadrature computation targeting reconfigurable hardware to maximize performance at a given error tolerance level. The authors proposed methodology that optimized performance by considering integration grid density versus mantissa size of floating-point operators. The optimization provided a number of integration points and mantissa size with maximized throughput while meeting given error tolerance requirement.

Three case studies showed that the proposed reduced precision designs on a Virtex-6 SX475T FPGA were up to 6 times faster than comparable FPGA designs with double-precision arithmetic. They were up to 15.1 times faster and 234.9 times more energy efficient than an i7-870 quad-core CPU and were 1.2 times faster and 42.2 times more energy efficient than a Tesla C2070 GPU.

2.2.2.5 Example 5: Customizable Architectures for the Set Covering Problem [Guo2013]

Classification #2 – Optimization toolbox
Subclassification – A particular single algorithm for a targeted set of applications
Applications/algorithms – NP-hard set covering problem

In this example, the authors proposed novel customizable streaming architectures for the NP-hard set covering problem (SCP). Both exhaustive and genetic algorithm approaches were covered, supporting coarse-grained parallelism and deep pipelines while allowing trade-offs between performance and resource usage.

The authors created streaming architectures for an exhaustive algorithm and a genetic algorithm that can be customized to support trade-offs between performance and resource usage. Their implementation of the FPGA is a good example of a trade-off between performance and quality of solutions. These FPGA implementations compared with existing FPGA designs showed improved flexibility and increased ability to support larger-scale genetic algorithms.

Experiments targeting Maxeler systems in this example showed that FPGA-based designs were more effective than the corresponding multi-core software versions. The speedups that were achieved exceeded 250 for the exhaustive algorithm and 60 for the genetic algorithm.

2.2.2.6 Example 6: A Fully Pipelined Probability Density Function Engine for Gaussian Copula Model [Ruan2014]

Classification #2 – Numerical analysis, modeling, and simulation toolbox
Subclassification – A particular single algorithm for a targeted set of applications
Applications/algorithms – Gaussian copula probability density function

In many fields where the multivariate dependence is of considerable interest, such as finance, hydrological modeling, biomedical study, and wavelet-based texture modeling, the Gaussian copula is a widely used multivariate modeling tool. However, existing solutions failed to achieve satisfactory performance, and users had to tolerate high computational complexity and the related time cost. This is because the mathematical model of the Gaussian copula, including either its CDF or the corresponding PDF, consists of plenty of time-consuming computation operations.

In this example, Xiaomeng Huang et al. developed an optimized FPGA-based Gaussian copula PDF evaluation scheme, which was able to achieve both high computation efficiency and low resource cost. It is a fast Gaussian copula PDF evaluation engine capable of handling all the time-consuming computation operations in a fully pipelined manner. Three optimization strategies were used during the process of deploying the originally CPU-friendly Gaussian copula PDF algorithm on the FPGA architecture:

- By transforming the calculation pattern of the Gaussian copula PDF algorithm, the authors significantly reduced the consumption of the computational resources.
- By eliminating constant computations from hardware logic, the authors achieved a better computational resource balance between the host and the accelerator.
- By extending calculations to multiple pipelines in one pass, the authors effectively exploited the utilization of the computational resources and achieved a significant performance improvement.

Overall, by applying the above optimizations, using one Virtex-6 SX475T FPGA, the authors achieved 1870 times speedup over a single-core CPU solution and 610 times speedup over a quad-core CPU solution. Furthermore, the performance of this solution can be easily scaled for future FPGA devices with more hardware resources since all instances processed by the Gaussian copula PDF are independent.

2.2.3 Examples from Exascale Science and Technology Drivers Leaf #1: Biology

2.2.3.1 Example 1: Hardware Acceleration of Genetic Sequence Alignment [Arram2013]

Classification #2 – Numerical analysis, modeling, and simulation toolbox
Subclassification – A particular single algorithm for a specific application
Applications/algorithms – DNA sequencing, alignment processor based on a backtracking variation of the FM-index algorithm

Next-generation DNA sequencing machines have been improving at an exceptional rate. The necessity to process so high quantities of data showed that the subsequent analysis of generated sequenced data had become a bottleneck in current systems. In this paper, J. Arram et al. explored the use of reconfigurable hardware to accelerate short read mapping problem, where the positions of millions of short DNA sequences are located relative to a known reference sequence.

The proposed design has an alignment processor based on a backtracking variation of the FM-index algorithm. Providing a full solution to the short read mapping problem, this design is capable of efficient exact and approximate alignment.

Nowadays, next-generation sequencing (NGS) machines are able inexpensively and at a very fast rate to produce sequenced data. In order to improve the throughput and measurement accuracy of these machines, the shorter sequences are processed. This allows tens of billions of bases to be sequenced per day. Short sequences are created by randomly breaking a long DNA chain. This random breaking causes the position and orientation information of the fragments with respect to the sample to be lost. Since there is the assumption that all DNA sequences within a species are similar, the sample DNA can be reconstructed by determining the location of the short fragments (the short reads) in a known reference genome of the species.

The authors created an application that involves highly parallel bit-oriented operations based on a backtracking FM-index algorithm, using FPGA technology as a good candidate for its acceleration. The design represents a full solution to short read mapping, capable of both approximate and exact alignment.

Essence
The major contributions of this work include:

– A hardware design, based on a backtracking FM-index algorithm, for a novel sequence alignment processor. Also, various optimizations, such as those for latency, memory size, and memory bandwidth, are analyzed.
– A Maxeler MAX3 board implementation of the proposed design.
– Performance evaluation done by comparing proposed design with some of the fastest software solutions on multi-core processors, GPUs, and FPGAs.

Infrastructure
The design uses a well-populated FPGA with alignment processors that is connected to the host processor using software driver. The software driver's role is, before the alignment starts, to transfer the Burrows-Wheeler transform (BWT) sequence to the accelerator board. BWT is then stored using on-chip BRAM or external DRAM. Short reads are inputted to the alignment processors in batches given by the design latency. Each batch of short reads is processed for a number of iterations. That number is determined by the permitted number of mismatches and by the short read length. The alignment results for each short read, including the SA interval, the cost, and a string representation of the alignment, are all reported to the software driver. This architecture is illustrated in Fig. 2.2.

The software driver has minimal role in the design architecture because the alignment processor design can be fully mapped to hardware. It is assumed that the BWT sequence is generated in advance because the reference sequence

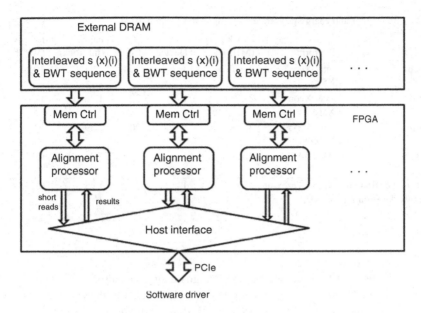

Fig. 2.2 Design architecture [Arram2013]

changes infrequently. The short reads are streamed to the alignment processors
populating the FPGA after transferring the data structures to the accelerator board
and configuring the number of mismatches permitted. The software driver then
waits until all the accelerator output is received. An SA interval is mapped to
positions in the reference sequence using a simple lookup table when a short read
can be matched to the reference sequence. This step can be performed in hardware;
however, it was chosen to perform it using a CPU in order to reduce the number
of memory controllers required by each alignment processor. When a short read is
unaligned, permitted mismatches are set to a higher number and the short read is
streamed again to the alignment processors for testing.

Relevance and Details
The authors proposed an alignment processor design with the following features:

– A novel scheme to reduce the memory size of the FM-index occurrence array,
  allowing it to be stored directly on the accelerator board
– A new method to reduce external memory access frequency, while maximizing
  memory bandwidth utilization
– A backtracking version of the FM-index algorithm with a data structure that
  supports both forward and backward search, which is capable of exact and
  approximate alignment
– A maximized throughput using a novel scheme to process batches of short reads
  in parallel

Table 2.3 Aligner and energy performance comparison [Arram2013]

Design	Platform	Clock freq. (MHz)	Devices	Cores	baps (millions)	Energy (W * hr)
Bowtie	Intel Xeon X5650	2,670	1	20	1.04	19
BWA	Intel Xeon X5650	2,670	1	20	1.76	11
SOAP2	Intel Xeon X5650	2,670	1	20	1.59	13
SOAP3	NVIDIA GTX 580	900	1	512	3.84	6.3
Proposed design on Max workstation (3 cores)	Xilinx Virtex-6 SX475T	150	1	3	13.5	0.078

Application and Results
Since it is difficult to directly compare designs using the raw results, the authors defined the bases aligned per second (baps) value as a normalized performance measure unit in order to better assess the performance of various designs.

Table 2.3 presents the achieved results.

Trends
Future research should include further optimization of this approach, together with its application in clinical procedures. It should provide that short reads with more than two mismatches increase the sensitivity of the software aligners from present (<20 %). This is a result of the aligner being unable to explore large search space within the cutoff time. The design's short reads with two mismatches had a comparable sensitivity (~100 %) to other software aligners.

2.2.3.2   Example 2: A Large-Scale Spiking Neural Network Accelerator for FPGA Systems [Cheung2012]

Classification #2 – Complex networks analysis toolbox
Subclassification – A particular single algorithm for a specific application
Applications/algorithms – Alignment processor based on a backtracking variation
    of the Ferragina-Manzini (FM) index algorithm

Spiking neural networks (SNN) aim to mimic membrane potential dynamics of biological neurons. They have been used widely in neuromorphic applications and neuroscience modeling studies. Although there is a lot of anatomical and functional knowledge of the brain, the scientists still don't have the complete picture of how higher cognitive function emerges from neuronal and synaptic dynamics. Large-scale simulation is useful in this regard, since it can be investigated how such functions emerge from deterministic simulation. The authors designed a parallel

SNN accelerator for producing large-scale cortical simulation targeting an off-the-shelf FPGA-based system. The accelerator parallelizes synaptic processing with run time proportional to the firing rate of the network.

Using only one FPGA, this accelerator is estimated to support simulation of 64 K neurons. The accelerator is 1.4 times (localized connectivity) to 5.5 times (uniform connectivity) faster than a GPU NeMo accelerator (Tesla C1060 65 nm process) in terms of spike delivery rate.

2.2.4 Examples from Exascale Science and Technology Drivers Leaf #2: Geophysics

2.2.4.1 Example 1: Maximum Performance Computing with DataFlow Engines [Pell2012]

Classification #2 – Optimization toolbox
Subclassification – A set of related algorithms composed as a complete stand-alone
    solution for a specific application
Applications/algorithms – Wave modeling application

In this article, Oliver Pell et al. discuss multidisciplinary DataFlow computing as a powerful approach to scientific computing that has led to orders-of-magnitude performance improvements for a wide range of applications. As an example, they provide one particular application – FD wave modeling.

One of the main factors that determine the resolution of seismic images is the bandwidth of the seismic wavelet. Finite-difference (FD) modeling and reverse-time migration (RTM) encounter particular problems increasing the wavelet bandwidth at the upper end of the spectrum because of the large impact this has on the computation resource requirements. Increasing the upper modeled frequency requires a finer spatial sampling, while the Courant Friedrichs Lewy (CFL) limit implies that the modeling time step must decrease. The amount of required computation increases with the fourth power of the wave frequency. It means that modeling at high frequencies (for example, 70Hz) can easily require hundreds of gigabytes of memory.

Essence
The CPUs in the system retain control of the application in DataFlow-accelerated wave modeling and instruct the DataFlow engine(s) to compute each time step. In each local memory of every DataFlow engine, pressure and velocity data volumes reside. They are streamed through the DataFlow implementation of the modeling kernel, which computes the next time-step pressure field. The data can also be read out from the DataFlow engine each time step to visualize or store to disk, while stimulus data is sent from the CPU to the DataFlow engine to be added to points in the field at run time.

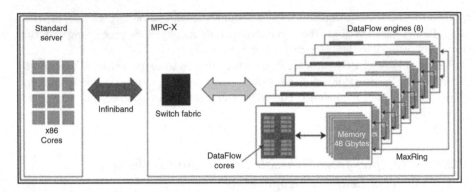

Fig. 2.3 MPC-X series architecture. These are stand-alone DataFlow compute nodes that connect to CPU nodes in a system via InfiniBand (Courtesy of Maxeler Technologies)

Infrastructure (Fig. 2.3)

Multiple DataFlow engines can work together on a single modeling problem, by splitting the domain in one dimension into multiple subdomains and assigning each subdomain to a different engine. At the edge of each subdomain, engines must exchange boundary data with their neighbors, and they can do this using the direct MaxRing interconnect. Since the MaxRing is a point-to-point interconnect, the communication bandwidth scales as the number of engines increases.

The Maxeler MAX2 FPGA card used has two Xilinx Virtex-5 FPGAs and 24GB of onboard DRAM. MAX2 has a high-speed MaxRing link between cards that enables multiple MAX2 cards to work together to achieve the 70Hz bandwidth objective.

Relevance and Details

Finite difference is the most commonly used numerical method for solving partial differential equations such as the wave equation. Explicit FD is an elegant and regular algorithm that affords efficient implementation within the DataFlow paradigm especially for the geosciences, medical imaging, and physics simulations.

Application and Results

The performance of the MAX2 system is compared to a C++ software version on a cluster with 32 3GHz X86 cores communicating via MPI over InfiniBand. The maximum performance of the accelerated node is equivalent to nearly 2000 CPU cores: one MAX2 card provides the equivalent performance of over 200 CPU cores. [Pell2014]

Characteristics
DataFlow computing can deliver orders-of-magnitude improvements in space and power consumption for a wide range of applications; and DataFlow compute engines can be balanced with other kinds of compute resources in a cluster environment (such as CPUs and storage). The benefits that can be seen show considerable promise for achieving the potential of exascale computing.

2.2.4.2 Example 2: An Implementation of the Acoustic Wave Equation on FPGAs [Nemeth2008]

Classification #2 – Numerical analysis, modeling, and simulation toolbox
Subclassification – A set of algorithms (framework) for a specific application
Applications/algorithms – The acoustic forward modeling application, 3D finite difference

FPGA chips are utilized as coprocessors in a PCI Express configuration to accelerate an acoustic isotropic modeling application. The acoustic forward modeling application in consideration is 3D finite difference, with 4th-order in time and 12th order in space, and uses single-precision floating-point arithmetic. The acoustic variable density modeling code contains a kernel which consumes majority of the compute cycles, indicating that the algorithm is a good candidate for acceleration. The finite-difference operators are calculated to minimize the relative phase velocity error over the bandwidth.

Optimization provides a peak speedup of over 160 times over a single core or 28–48 times speedup per node depending on multi-core scaling. It can be observed that FPGA speedup increases further for problem sizes beyond 400 mesh, too.

2.2.4.3 Example 3: Finding Speedup in Parallel Processors [Flyn2008]

Classification #2 – Numerical analysis, modeling, and simulation toolbox
Subclassification – A set of related algorithms composed as a complete stand-alone solution for a targeted set of applications
Applications/algorithms – Geophysical modeling, forward modeling a finite-difference method

One particularly compute-intensive application is concerned with oil and gas exploration. In this application, data is collected by first distributing a grid of sensors over a large area. Then a sonic impulse is applied to the area and reflections are recorded: frequency, amplitude, and delay at each sensor. Sonic impulse could be a compressed air cannon (at sea) or explosives (on land).

A typical sea-based survey uses 30,000 sensors to record data (over a 120 dB dynamic range). With a new sonic impulse occurring every 10 sec, each sensor was sampled at more than 2 kbps. Reflections of these impulses that the earth structures reflect are detected by a sensor array. In this way, terabytes of resulting data are created each day.

In this article, the authors propose an acceleration methodology based on FPGA arrays. The methodology uses a comprehensive application analysis supported by high-performance FPGA hardware. The analysis provides a DataFlow graph of the application which is replicated in SIMD for multiple data strips (until limited by the pin bandwidth), then pipelined (MISD) until circuit limited.

In this particular application, the FPGA solution shows the possibility of speedup of over 300 times over an Intel Xeon solution.

2.2.4.4   Example 4: Anisotropic Reverse-Time Migration Using Co-Processors [Liu2009]

Classification #2 – Image, video, text processing, and analysis toolbox
Subclassification – A particular single algorithm for a targeted set of applications
Applications/algorithms – Seismic imaging, anisotropic reverse-time migration

Coprocessors offer attractive acceleration opportunities to waveform-based imaging and inversion applications in challenging exploration and production environments. Unlike seismic forward modeling, the large amount of data involved in seismic imaging and inversion can pose a significant challenge to scalable acceleration. The authors provide and compare several computational schemes to perform anisotropic reverse-time migration on two coprocessor platforms: FPGAs and GPUs. The ongoing experiments so far indicate that both platforms can potentially achieve high speedups using acceleration-friendly schemes which minimize interruptions to computation from data movement and storage.

FPGA-accelerated isotropic modeling reached a speedup rate of 20 times over 8 cores. Currently, accelerated isotropic wave propagation on FPGAs can go beyond 40 times (over 8 cores) or 200 times over a single core.

2.2.5   Examples from Exascale Science and Technology Drivers Leaf #3: Meteorology

2.2.5.1   Example 1: Acceleration of a Meteorological Limited Area Model with DataFlow Engines [Oriato2012]

Classification #2 – Machine learning and data mining toolbox
Subclassification – A set of algorithms (framework) for a specific application
Applications/algorithms – Hydrostatic limited area model derived from the BOLAM model

Due to hard deadlines inherent in predicting weather, climate and weather modeling needs High Performance Computing. This modeling produces large data volumes so it is ideally suited for DataFlow computation. In this paper, Diego Oriato et al. demonstrate a DataFlow implementation of dynamic core of a meteorological limited area model. The authors focused on dynamic core of a limited area model (LAM) derived from the BOLAM model. It is a research-oriented hydrostatic LAM developed by ISAC-CNR (Bologna, Italy) and parallelized using message passing libraries and domain decomposition [Marrocu1998].

They presented results for a domain of $13,600 \times 3,333 \times 30$ km with 620 thousand grid points. In order to satisfy requirements of higher spatial resolution for a regional weather forecast, large domain dimension for global models, and very long time integration typical of climate simulations, it was necessary to develop a computationally fast version of the dynamic core of the hydrostatic model.

Essence

The paper gives an example of a possible standard procedure that could be taken in transferring application from a control flow computer to a Maxeler DataFlow engine. A complete version of a hydrostatic LAM is made up of three main units: initialization, post-processing, and the meteorological model that numerically solves the prognostic equations governing atmospheric circulation. The last part is composed of two macro blocks: the dynamic core and the physical parameterization routines. This part is the most complex and computationally expensive (especially the radiation). The primitive equations are a set of nonlinear 3D partial differential equations that approximate global atmospheric flow. They consist of equations for the conservation of thermal energy, continuity, and momentum.

The whole work was divided in four phases.

- In *Structural Analysis* phase, time step computation was decomposed into five logical blocks. Each block is structured as a 3D loop where longitude, latitude, and altitude are, respectively, fast, medium, and slow dimensions.
- In *Partitioning* phase, the authors used Maxeler Parton toolset to analyze the CPU time spent on each logical block of the application when run on an Intel Xeon core. The toolset gave as a result a lot of relevant data. The authors decided to migrate all the blocks to DFE because an acceleration of 100 times or more was the goal. By moving all the computation to DFE, data transfer between CPU and DFE was also minimized.
- In *Transformation* phase, DataFlow computing operations were implemented spatially as part of a pipeline through which data was streamed rather than each instruction being executed temporally on a new piece of data. For this reason, a DataFlow algorithm is inherently parallel. The aim was therefore to group all the computational logic inside a single loop and to replace the loop with a deterministic data access pattern. This required an understanding of data dependency among different parts of the algorithm.

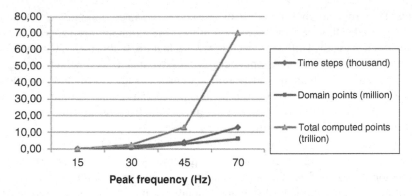

Fig. 2.4 Impact of increased modeling frequency on memory and computation costs. In wave modeling, the necessary amount of computation increases with the fourth power of the wave frequency [Pell2012]

– In *Parallelization* phase, an analysis was done how in detail to parallelize the algorithm and the DataFlow. A DataFlow engine comprises memory coupled to a chip implementing many DataFlow cores, as shown in Fig. 2.4. The DataFlow cores are arranged in a pipeline which processes one item of data per clock cycle. To facilitate parallelization part, the computation was converted to use fixed-point arithmetic. Namely, fixed-point arithmetic is more efficient in terms of silicon area per computer operation than the equivalent floating-point arithmetic. Analysis was performed on each block of the algorithm to establish the optimal fixed-point representations to maximize precision and avoid overflow, applying scaling coefficients where prognostic variables exhibited high dynamic range. The authors also chose to use one-to-one mapping of serial simulation per DFE, giving the capability of running six independent simulations in parallel.

Infrastructure
The DataFlow application was run on a Maxeler MPC-C series DataFlow node (eight Intel Xeon E5506@2.13GHz CPU cores and six MAX3 DataFlow engines connected to the CPUs via PCI Express). Each MAX3 DFE utilizes a Xilinx Virtex-6 SX475T FPGA to implement DataFlow cores and 48GB of memory. Maxeler's MaxCompiler development environment was used to implement the DataFlow pipeline and integrate it into the original FORTRAN application.

No computation was executed on the CPU other than for initialization and post-processing. All the variables needed during the computation were stored on the DFE. Only the prognostic variables were initialized from CPU and transferred out at the last time step of the simulation. The CPU controlled the time step loop by triggering the DataFlow process and waiting for it to finish before repeating the trigger.

Application and Results
The speedups acquired by using a single DFE was 64 times compared to a single-node single-core CPU and 12 times compared to a single-node 12-core CPU. When MPC-C series node was used, the speedup was 381 times compared to a single-node single-core CPU and 74 times compared to a single-node 12-core CPU.

Peak power usage of a MPC-C series node was measured to be around 900 W. Considering a peak value of 400 W per Intel node and 74 twelve-core nodes, around 30 kW would be needed to match the performance of the DataFlow solution.

Characteristics
This is an example of how a good preparation and analysis done prior to the migration process of an application can give really good results, not only in speedup but also in power reduction.

2.2.6 Examples from Exascale Engineering and Innovation Drivers Leaf #5: Oil Industry

2.2.6.1 Example 1: Surviving the End of Frequency Scaling with Reconfigurable DataFlow Computing [Pell2011]

Classification #2 – Optimization toolbox
Subclassification – A set of algorithms (framework) for a specific application
Applications/algorithms – Common Reflection Surface (CRS) seismic trace stacking, a fitness function

In this paper, besides presenting a very good example of a DFE usage in the oil and gas industry, Oliver Pell et al. give an excellent overview of Maxeler FPGA computer, of how the heterogeneous computing was the solution for frequency crunch in CPU/GPU systems, how parallelism was built and could be used in multi- and many-core systems (CPU/GPU), and finally how the only real available way to exascale supercomputing that wasn't at the same time devouring consumers of electrical power was to use and to take advantage of capabilities of both the worlds – control flow and DataFlow computers.

It is by no means suggested that DataFlow engines were suitable to replace the conventional CPU entirely. They should be used to increase the performance of control flow processors. Most of the lines of code in a program will still run on the CPU. Only computationally intensive components of an application should be offloaded to a custom DataFlow engine (Fig. 2.5). A typical Maxeler HPC compute node consists of some number of CPUs coupled to some number of FPGAs; for example, one current standard 1U node has 12 Intel Xeon CPU cores, 4 Xilinx Virtex-6 FPGAs, and 100–400GB of RAM split between the CPUs and FPGAs. Large RAM capacities are important for many of the target applications, which process very large data volumes.

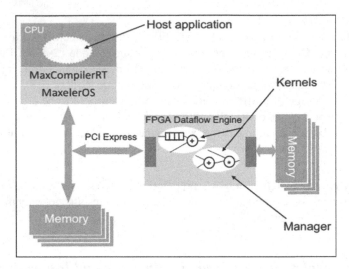

Fig. 2.5   A CPU coupled to a custom FPGA DataFlow engine (Courtesy of Maxeler Technologies)

The main challenge in the practical exploitation of DataFlow engines is to program them. Unlike writing code for a CPU where the programmer is creating a set of instructions that will be executed by existing function units, to create a DataFlow engine, the programmer must actually construct a circuit that represents the application. However, this does not need to be an electrical engineering process of circuit design – Maxeler has developed a programming tool called MaxCompilerRT that allows software engineers to implement DataFlow engines for an application using a high-level software environment.

Essence

The oil and gas industry is one of the major industry consumers of HPC computing. The world has been exploiting reserves of "the black gold" for more than 150 years, and after depleting the existing reservoirs, the new ones have to be found. In order to find them, the scientists need to make images of the subsurface. To create those images, scientists perform acoustic experiments on the earth's surface. A low-frequency source is activated and the reflections from the different subsurface layers are recorded by thousands of sensors. This experiment is repeated many times, and it creates hundreds of terabytes of data. To process this data, thousands of compute nodes are used and the processing lasts a very long time.

Infrastructure

CRS stacking is an algorithm used to process seismic survey data to compute zero-offset traces. CRS equation needs input data given in 8 parameters to compute zero-offset traces. Before it can compute the stack, the stacking application must

Fig. 2.6 Percentage of
computation time taken in
each function in the original
software [Pell2011]

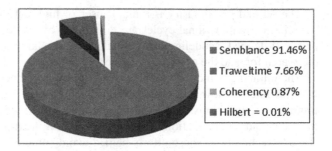

determine good values for the eight parameters by performing a search in 8-dimensional space. A fitness function is evaluated at each point in the space to determine the quality of the current parameter set.

The control flow architecture execution time to compute CRS for a typical survey can be in the order of 1 month using 1000 CPU cores. As presented in Fig. 2.6, this runtime is predominately spent on the computation of the Semblance (fitness) and Traveltime. The semblance function computes on a window of data values from that location, while the travel time function calculates the location that should be read from a data trace based on the eight CRS parameters.

Application and Results
With both semblance and traveltime computation done on the FPGA, over 99 % of the total run time from the CPU was accelerated. The final implementation gave a speedup of approximately 230 times compared to a single core for land datasets and 190 times for marine datasets, what was approximately 30 times greater performance/Watt.

Characteristics
Definitely a lot of sciences and industries can successfully use heterogeneous computing architecture. The best results, up to now, have been achieved in the fields of finances and oil and gas industries. This is an example of how huge speedups such architecture can offer.

2.2.6.2   Example 2: Beyond Traditional Microprocessors for Geoscience High-Performance Computing Applications [Lindtjorn2011]

Classification #2 – Optimization toolbox
Subclassification – A set of algorithms (framework) for a specific application
Applications/algorithms – Reverse-time migration

The oil and gas industry is a major user of high-performance computing, and geoscience computational cycles are dominated by kernels that are relatively few and well defined. Modeling and computation are taking an unprecedented role in the search for, and the extraction of, energy sources like oil and gas.

The oil and gas industry already uses high-performance computing (HPC), but it's still unclear how conventional HPC technologies can meet the demands of tomorrow's algorithms. In this article, authors described the acceleration of the most demanding applications in this domain using field-programmable gate array (FPGA) technology. These efforts helped avoid some of the performance-scaling issues frequently encountered with CPUs and GPUs.

This approach delivered speedups of 20–70 times compared to a conventional HPC node.

2.2.6.3   Example 3: Acceleration of Anisotropic Phase Shift Plus Interpolation with DataFlow Engines [Tomas2012]

Classification #2 – Optimization toolbox
Subclassification – A set of algorithms (framework) for a specific application
Applications/algorithms – Phase Shift Plus Interpolation method

Although time-domain depth migration techniques have been successfully ported to run on modern hardware accelerators, their ultimate obstacle is the I/O overhead present during the imaging step. Frequency-domain depth migration algorithms overcome this limitation and can exploit the full potential of new computing technologies. In particular, this implementation of Phase Shift Plus Interpolation (PSPI) method is characterized by fast running time, good-quality results under low signal-to-noise ratio conditions, and excellent results for steep dips.

The measurements indicated that a DataFlow approach could achieve high speedups despite larger and larger computational domains, increased complexity of the anisotropic approach, and the I/O overhead during angle-gathers calculation. When MAX2 and MAX3 systems were compared to a FORTRAN software version on a node with 8 Intel Xeon 2.6GHz cores parallelized using MPI, the speedup was from 13 to 34 times.

2.2.6.4   Example 4: Accelerating Large-Scale HPC Applications Using FPGAs [Dimond2011]

Classification #2 – Optimization toolbox
Subclassification – A particular single algorithm for a targeted set of applications
Applications/algorithms – Wave propagation using 3D finite difference

The key to achieving the best performance in FPGA accelerators, while maintaining accuracy, is optimization of arithmetic units and data types to suit the range/precision at each point in the computation. The flexibility of FPGA to implement nonstandard arithmetic, combined with a DataFlow programming model that instantiates a separate unit for each arithmetic operator in the code, provides a wide design space. As such, FPGA computing offers significant opportunity for arithmetic research into "large-scale" HPC applications, where there is a chance to move away from standard IEEE formats, either to improve precision compared to the CPU version or to increase speed.

The key to performance in stacking on the FPGA is to maximize reuse of each point loaded from DRAM; otherwise, the available DRAM bandwidth into the chip is the limit to performance. The degree of reuse of each input data point changes with varying the number of output data points computed in parallel. When doing 64 parallel searches, the degree of reuse of each input data point, while varying the number of output data points computed in parallel, is high enough so that the application is no longer memory bound.

The authors used the flexibility of FPGA arithmetic to trade off precision and performance in different versions of the application. A full-precision version, used for pricing accurate to 10^8, gave a 31 times speedup over an 8-core Xeon E5430 server.

2.2.6.5 Example 5: Accelerating 3D Convolution Using Streaming Architectures on FPGAs [Fu2009]

Classification #2 – Optimization toolbox
Subclassification – A set of algorithms (framework) for a specific application
Applications/algorithms – 3D convolution

In this paper, Haohuan Fu et al. investigate FPGA architectures for accelerating applications whose dominant cost is 3D convolution, such as modeling and reverse-time migration (RTM). The authors explore different design options, such as using different stencils, fitting multiple stencil operators into the FPGA, processing multiple time steps in one pass, and customizing the computation precisions. The exploration reveals constraints and trade-offs between different design parameters and metrics.

They are using two major FPGA advantages over other computation platforms: (1) streaming computation architecture and (2) customizable number representations. They experimented with processing different numbers of time steps in one pass and got different speedups, and the same happened when different floating-point precisions were used.

The experiment results showed that the FPGA streaming architecture provided great potential for accelerating 3D convolution and could achieve a speedup of up to two orders of magnitude. By dividing the array into two parts and computing in two FPGAs concurrently, speedups of up to 55 times and up to 47 times were achieved.

2.2.6.6 Example 6: Finite-Difference Wave Propagation Modeling on Special-Purpose DataFlow Machines [Pell2013]

Classification #2 – Numerical analysis, modeling, and simulation toolbox
Subclassification – A particular single algorithm for a targeted set of applications
Applications/algorithms – 3D finite-difference calculations

Modeling wave propagation through the earth is an important application in geoscience. In this paper, Oliver Pell et al. present a framework for wave propagation modeling on special-purpose hardware, which dramatically improves the application performance compared to conventional CPUs. They use custom hardware platforms consisting of a mix of x86 CPUs and DataFlow engines connected by high bandwidth communication links.

The application-specific DataFlow engines run at hundreds of MHz with massive parallelism and deliver high performance/Watt, meaning that they are up to 30 times more energy efficient than conventional CPUs. The power efficiency of this approach suggests that DataFlow computing may have a key role to play in the improvements in power efficiency necessary to reach exascale computing.

2.2.7 Additional Research Papers

In the last couple of years, many postgraduate students in countries of Southern Europe have been doing intensive experimenting with porting algorithms and applications to Maxeler DataFlow engines. A few of their research papers were published in a special July 2013 issue "Maxeler Super Computer Related Research" of *IPSI Transaction on Internet Research* magazine.

The examples of Exascale Fundamental Drivers: Mathematics (leaf #2) can be found in [Stanojevic2013], [Rankovic2013], [Bezanic2013], and [Sustran2013].

The example of Exascale Science and Technology Drivers: Meteorology (leaf #3) can be found in [Ivkovic2013].

The examples of Exascale Science and Technology Drivers: Physics (leaf #4) can be found in [Stojanovic2013] and [Korolija2013].

References

[Arram2013]   Arram J et al (2013) Hardware acceleration of genetic sequence alignment. In: Proceedings of the 9th international symposium on reconfigurable computing: architectures, tools and applications, Los Angeles, CA, 2013. pp 13–24
[Bezanic2013]   Bezanic N et al (2013) Implementation of the RSA algorithm on a dataflow architecture. IPSI Trans Intern Res Belgrade Serbia 9(2):11–18

[Chow2012]   Chow GCT et al (2012) A mixed precision Monte Carlo methodology for reconfigurable accelerator systems. In: Proceedings of the ACM/SIGDA international symposium on Field Programmable Gate Arrays (FPGA), Monterey, CA, 2012, pp 57–66

[Chau2013]   Chau TCP et al. (2013) Heterogeneous reconfigurable system for adaptive particle filters in real-time applications. In: Proceedings of the 9th international symposium on reconfigurable computing: architectures, tools and applications, Los Angeles, CA, 2013, pp 1–12

[Cheung2012]   Cheung KS, Schultz R, Luk W (2012) A large-scale spiking neural network accelerator for FPGA systems. In: Proceedings of the 2nd international conference on artificial neural networks, Lausanne, Switzerland, 2012, Part I, pp 113–120

[Dimond2011]   Dimond D, Racaniere S, Pell O (2011) Accelerating large-scale HPC applications using FPGAs. In: Proceedings of the 20th IEEE symposium on computer arithmetic (ARITH), Tubingen, Germany, 2011, pp 191–192.

[Flynn2008]   Flynn M et al (2008) Finding speedup in parallel processors. In: Proceedings of the international symposium on parallel and distributed computing, Krakow, Poland, 2008, pp 3–7

[Fu2009]   Fu H, Clapp RG, Mencer O, Pell O (2009) Accelerating 3D convolution using streaming architectures on FPGAs. In: SEG Houston 2009 international exposition and annual meeting, Houston, Texas, 2009.

[Guo2012]   Guo C, Fu H, Luk W (2012) A fully-pipelined expectation-maximization engine for Gaussian mixture models. In: Proceedings of the 2012 international conference on field-programmable technology (FPT), Seoul, S. Korea, 2012, pp 182–189

[Guo2013]   Guo L et al (2013) Customizable architectures for the set covering problem. Comput Architect News ACM Sigarch, New York 41(5):101–106

[Ivkovic2013]   Ivkovic S et al (2013) Source-sink model. IPSI Trans Inter Res Belgrade Serbia 9(2):28–33

[Jin2012]   Jin Q et al. (2012) Multi-level customization framework for curve based Monte Carlo financial simulations. In: Proc. 8th international conference on reconfigurable computing: architectures, tools and applications, Hong Kong, 2012, pp 187–201

[Korolija2013] Korolija N et al (2013) Accelerating Lattice-Boltzmann method using the Maxeler DataFlow approach. IPSI Trans Internet Res Belgrade Serbia 9(2):34–42

[Kumar2009]   Kumar N, Satoor S,, Buck I (2009) Fast parallel expectation maximization for Gaussian mixture models on GPUs using CUDA. In: EEE international conference on high performance computing and communications, 2009, pp 103–109

[Lindtjorn2011] Lindtjorn O et al (2011) Beyond traditional microprocessors for geoscience high-performance computing applications. Micro IEEE 31(2):41–49

[Liu2009]   Liu W et al (2009) Anisotropic reverse-time migration using co-processors. SEG Houston 2009 international exposition and annual meeting, Houston, Texas, 2009

[Marrocu1998]   Marrocu M, Scardovelli R, Malguzzi P (1998) Parallelization and Performance of a Meteorological Limited Area Model. Parallel Comput 24(5–6):911–922

[Mencer2011]   Mencer O et al (2011) Finding the Right Level of Abstraction for Minimizing Operational Expenditure. iN: Proceedings of the 4th workshop on high performance computational finance, ACM New York, NY, 2011, pp 13–18

[Nemeth2008]   Nemeth T et al. (2008) An implementation of the acoustic wave equation on FPGAs. In: Proceedings of the 78th Society of Exploration Geophysicists (SEG) meeting, Las Vegas, November 2008, pp 2874–2878

[Oriato2012]   Oriato D et al (2012) Acceleration of a meteorological limited area model
                    with dataflow engines. In: Proceedings of the 2012 symposium on application
                    accelerators in high performance computing, Chicago, IL, July, 2012, pp 129–
                    132
[Pell2011]   Pell O, Mencer O (2011) Surviving the end of frequency scaling with recon-
                    figurable dataflow computing. Newsletter ACM SIGARCH Comput Architect
                    News Arch 39(4):60–65. ACM, New York, NY
[Pell2012]   Pell O, Averbukh V (2012) Maximum performance computing with dataflow
                    engines. Comput Sci Eng 14(4):98–103, Los Alamitos, CA
[Pell2013]   Pell O et al (2013) Finite-difference wave propagation modeling on special-
                    purpose dataflow machines. IEEE Trans Parallel Distrib Syst 24(5):906–915
[Pell2014]   Pell O et al (2014) Summary FD modeling beyond 70Hz with FPGA
                    acceleration [Online]. Available: http://www.maxeler.com/media/documents/
                    MaxelerSummaryFDModelingBeyond70Hz.pdf
[Rafique2012]   Rafique A, Kapre N, Constantinides GA (2012) Enhancing performance of
                    Tall-Skinny Qr Factorization using FPGAs. In: Proceedings of the 22nd
                    international conference on Field Programmable Logic and applications (FPL),
                    Oslo, Norway, August 2012, pp 443–450
[Rankovic2013]   Rankovic V, Kos A, Milutinovic V (2013) Bitonic merge sort implementation
                    on the Maxeler Dataflow supercomputing system. IPSI Trans Internet Res
                    Belgrade Serbia 9(2):5–9
[Ruan2014]   Ruan H et al (2014) A fully pipelined probability density function engine for
                    Gaussian Copula model. Tsinghua Sci Technol Tsinghua Univ Press (TUP)
                    19(2):194–202
[Stojanovic2013]   Stojanovic S, Bojic D, Milutinovic V (2013) Solving Gross Pitaevskii equation
                    using DataFlow paradigm. IPSI Trans Internet Res Belgrade Serbia 9(2):19–22
[Stanojevic2013]   Stanojevic I, Senk V, Milutinovic V (2013) Application of Maxeler Dataflow
                    supercomputing to spherical code design. IPSI Trans Internet Res Belgrade
                    Serbia 9(2):1–4
[Sustran2013]   Sustran Z, Todorovic M, Milutinovic V (2013) Feasibility study on the SAT
                    solver on DataFlow architecture. IPSI Trans Internet Res Belgrade Serbia
                    9(2):23–27
[Tomas2012]   Tomas C et al. (2012) Acceleration of anisotropic phase shift plus interpolation
                    with dataflow engines. In: Proceedings of the 82nd annual meeting and
                    international exposition of the Society of Exploration Geophysics-SEG, Las
                    Vegas, NE, 2012, pp 3402–3406
[Tse2012]   Tse HT (Anson) (2012) Accelerating reconfigurable financial computing. PhD
                    thesis, Imperial College, London, G. Britain, 2012
[Tse2012/2]   Tse AHT et al (2012) Optimizing performance of quadrature methods with
                    reduced precision. In: Proceedings of the 8th international symposium ARC
                    2012, Hong Kong, China, March, 2012, pp 251–263
[Weston2010]   Weston S et al (2010) Accelerating the computation of portfolios of Tranched
                    credit derivatives. In: Proceedings of the 2010 IEEE Workshop on High
                    Performance Computational Finance (WHPCF), New Orleans, LA, November
                    2010, pp 1–8
[Weston2011]   Weston S et al (2012) Rapid computation of value and risk for derivatives
                    portfolios. Concurr Comput Pract Exp 24(8):880–894

Chapter 3
An Example Application: Fourier Transform

3.1 Introduction

In this chapter we are going to show the process of accelerating an application and measuring achieved acceleration on one simple example. The example used in this chapter is acceleration of the Cooley-Tukey algorithm implemented in C++.

3.1.1 About the Cooley-Tukey Algorithm

The Cooley-Tukey algorithm is the most commonly used algorithm for calculating the fast Fourier transform. The algorithm was published in 1969 by J. W. Cooley and John Tukey in the paper "The Fast Fourier Transform and Its Applications" [Cooley1969]. The algorithm is based on divide and conquer design paradigm and works by calculating the discrete Fourier transform of the entire input sequence using the discrete Fourier transforms of the subsequences of the input sequence.

3.1.2 About the Fast Fourier Transform Algorithm

Algorithms for calculating the fast Fourier transform are used to calculate the discrete Fourier transform of a given input sequence. Discrete Fourier transform transforms discrete input signal from the time domain to the frequency domain. Fast Fourier transform is used in many different areas including digital signal processing, telecommunications, and sound signal analysis.

Time complexity of the fast Fourier transform algorithms on standard microprocessor systems, with the John von Neumann architecture, is $O(N \log N)$.

© Springer International Publishing Switzerland 2015
V. Milutinović et al., *Guide to DataFlow Supercomputing*, Computer Communications and Networks, DOI 10.1007/978-3-319-16229-4_3

The accelerated algorithm shown in this chapter works in time complexity $O(\log N)$. Although time complexity of the accelerated algorithm is much lower than the time complexity of standard algorithms, the reader should be aware of the fact that accelerating applications using Maxeler DFE technology introduces some differences, when compared to traditional John von Neumann programming model. On one side, the reader should know that when using Maxeler DFE to accelerate an application, some time is needed to transfer the data to and from the DFE. On the other hand, Maxeler DFE systems have the advantage that they could generate results of the computation on each clock cycle, after the period of initial latency, by making a hardware pipeline. This characteristic allows suitable applications to achieve high speedups when accelerated using a Maxeler DFE system.

3.1.3 Overview of Different Fast Fourier Transform Algorithms

Other than the Cooley-Tukey algorithm, there are many other algorithms for calculating the fast Fourier transform (FFT), like the prime-factor FFT algorithm, Bruun's FFT algorithm, Reader's FFT algorithm, Winograd's FFT algorithm, and Bluestein's FFT algorithm. These algorithms use different mathematical methods to calculate the fast Fourier transform. Used mathematical methods vary from number theory via numerical mathematics to graph theory. All these algorithms have in common that their time complexity is $O(N \log N)$ when running on a CPU. Although there is no evidence that it is impossible to construct a faster algorithm than the one with the time complexity $O(N \log N)$, faster algorithm has not yet been constructed.

Performance comparison of a large number of publicly available algorithms used to calculate the fast Fourier transformation was done by Matteo Frigo and Steven G. Johnson [Frigo2014] from Massachusetts Institute of Technology, USA. In their experiments, they used the implementations of various authors, written over a time period longer than 35 years. Experiments were executed on different types of computer architectures. It should be noted that Frigo and Johnson, in their experiments, used only one processor core in multiprocessor systems. The results of their experiments are publicly available and will be used in this chapter to compare the performance obtained using the Maxeler MAX3 DFE System with other publicly available implementations. Figure 3.1 is a graphical representation of data from one of Frigo's and Johnson's experiments.

Figure 3.1 shows a comparison between performances of different algorithms for calculating the fast Fourier transform of complex input sequences in single precision depending on the length of the input sequence. The experiment was carried out on the Intel Xeon 3.60 GHz Pentium 4 (Prescott) microprocessor. Scale of the Y axis, which measures the speed of the algorithm, is expressed in MFlops. MFlops in this

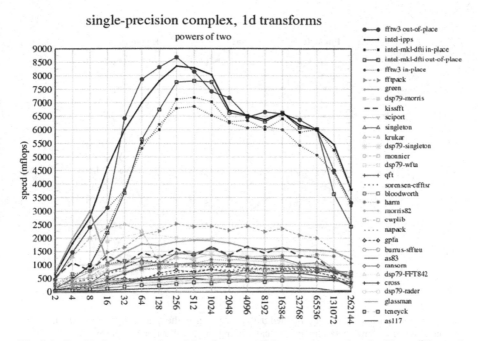

Fig. 3.1 Graphical representation of data from one of the experiments done by Matteo Frigo and Steven G. Johnson

case do not represent the number of floating point operations per second; instead MFlops value is obtained from the average execution speed of the algorithm using the following formula:

$$\text{mflops} = 5\,N\log_2(N)/\,(\text{time for one FFT in microseconds})$$

(Source: http://www.fftw.org/speed/Pentium4-3.60GHz-icc)

3.2   The Radix 2 Cooley-Tukey Algorithm

The radix 2 Cooley-Tukey algorithm consists of the following steps:

Step 1   Input sequence is divided into two equal subsequences. Elements of the input sequence with even indexes form first subsequence and elements with odd indexes form second subsequence.

Step 2   Calculate discrete Fourier transform for the two subsequences.

Step 3   Based on the discrete Fourier transform of the subsequences, calculate the discrete Fourier transform of the entire input sequence (Fig. 3.2).

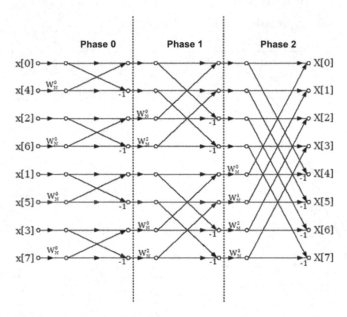

Fig. 3.2 Graphical illustration of pipelined data processing in the radix 2 algorithm. The illustration on this figure shows data pipeline for calculating the fast Fourier transform using the radix 2 algorithm. Input sequence shown in the illustration has eight points. Pipeline consists of three phases (Phase 0, Phase 1, and Phase 2). Number of stages is $\log_2(N)$, where N is the number of points in the input sequence

3.3   Mathematical Background

Discrete Fourier transform of the entire input sequence can be calculated from the discrete Fourier transform of even and odd subsequences in the following way:

$$X_k = \sum_{n=0}^{N-1} x_n e^{-\frac{2\pi i}{N} nk}$$

$$= \sum_{m=0}^{N/2-1} x_{2m} e^{-\frac{2\pi i}{N}(2m)k} + \sum_{m=0}^{N/2-1} x_{2m+1} e^{-\frac{2\pi i}{N}(2m+1)k}$$

$$= \sum_{m=0}^{N/2-1} x_{2m} e^{-\frac{2\pi i}{N/2} mk} + e^{-\frac{2\pi i}{N} k} \sum_{m=0}^{N/2-1} x_{2m+1} e^{-\frac{2\pi i}{N/2} mk}$$

where k is in range from 0 to $n-1$.

First sum in the last expression represents the discrete Fourier transform of the even subsequence, while second sum represents the discrete Fourier transform of the odd subsequence. If we mark the first sum as E_k, second sum as O_k, and $e^{-2\pi i/N}$ as W_N^k, we get the following formula:

$$X_k = E_k + W_N^k O_k$$

Because of the periodicity properties of the discrete Fourier transform, the following two equations are true:

$$E_{k+\frac{N}{2}} = E_k$$

$$O_{k+\frac{N}{2}} = O_k$$

Because of the properties of the Euler formula, the following two equations are true:

$$e^{-\frac{2\pi i}{N}(k+N/2)} = -e^{-\frac{2\pi i}{N}k}$$

$$W_N^k = -W_N^k$$

Combining previous equations with the formula for calculating the discrete Fourier transform we get the final formula that allows us to implement algorithm that we will use to calculate the discrete Fourier transform of the input sequence from it's even and odd subsequences.

$$X_k = E_k + W_N^k O_k, \quad k = 0\ldots N/2$$

$$X_{k+\frac{N}{2}} = E_k - W_N^k O_k, \quad k = 0\ldots N/2$$

3.4 Pseudo Code of the Radix 2 Cooley-Tukey Algorithm

Based on the formula derived in the last chapter, it is easy to come up with the following recursive algorithm for calculating discrete Fourier transform.

```
: calculates the discrete Fourier transform of input sequence x
X0, ..., N-1 ← fft_radix2(x):
    ; special case for the input sequence of length one
    if N = 1 then
        X0 ← x0
    else
        ; discrete Fourier transform of the even subsequence
        X0, ..., N/2-1 ← fft_radix2(even_elements(x))
        ; discrete Fourier transform of the odd subsequence
X N/2, ..., N-1 ← fft_radix2(odd_elements(x))
; calculating discrete Fourier transform from
; the discrete Fourier transforms of even and odd subsequences
        for k = 0 to N/2-1
            t ← Xk
            Xk ← t + exp(-2πi k/N) Xk+N/2
            Xk+N/2 ← t - exp(-2πi k/N) Xk+N/2
        endfor
    endif
```

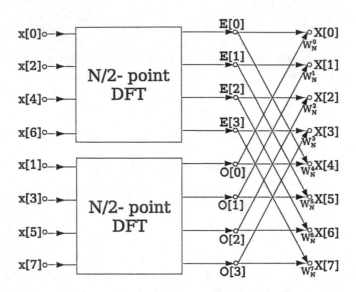

Fig. 3.3 Graphical illustration of the radix 2 algorithm. This illustration shows how the radix 2 algorithm works for the input sequence of eight points. Input sequence is divided into two subsequences. The first subsequence consists of the elements with even indexes; the second subsequence consists of the elements with odd indexes. Illustration shows how the discrete Fourier transform of the entire sequence is calculated from the discrete Fourier transforms of its subsequences

The above pseudo code is for the radix 2 Cooley-Tukey algorithm (Fig. 3.3).

3.5   Original Implementation

Original implementation of the Cooley-Tukey algorithm is iterative in-place radix 2 implementation. The implementation has optimal time complexity of $O(N \log N)$, where N is the number of points in the input sequence. Space complexity of the original implementation is $O(1)$. Code of the original implementation is written in C++ and is given in Sect. 3.12 of this book.

3.5.1   In-Place Implementation of the Cooley-Tukey Algorithm

In-place algorithms have space complexity $O(1)$. Constant memory consumption is achieved by overriding values of the input sequence with intermediate and final results. To implement Cooley-Tukey algorithm, in-place input or output sequence needs to be rearranged in bit reverse order.

Bit reverse order of original sequence is generated by swapping the place of elements that have reverse indexes to each other in binary format, for example, bit reverse order of sequence 0, 1, 2, 3, 4, 5, 6, 7 is 0, 2, 4, 6, 1, 3, 5, 7.

The following snippet of code reorders the sequence in bit reverse order:

```
for (i = 0; i < n; i++) {
        j = bit_reverse(i);

        if (i > j) {
                swap(real[i],  real[j]);
                swap(imag[i], imag[j]);
        }
}
```

3.6 Analysis of the Accelerated Version of the Cooley-Tukey Algorithm

Similar to the original CPU implementation of the Cooley-Tukey algorithm, accelerated version consists of rearranging the order of elements in input sequence to bit reverse order and the radix 2 algorithm.

MaxJ code of the kernel for calculating the fast Fourier transform using the radix 2 algorithm is given in the Sect. 3.12 of this book (Figs. 3.4 and 3.5).

3.7 Implementational Details

The code of the kernel for calculating the fast Fourier transform accepts parameter N that represents the number of points in the input sequence. Based on this parameter the end bitstream design, design that is going to be used on FPGA chip, is generated for calculating the fast Fourier transform.

This kernel has two input and two output data streams. These two streams are used for transferring real and imaginary parts of the input sequence points to and from the DFE. These streams are of type arrayType that is defined in the following way:

```
DFEType floatType = dfeFloat(8, 24);
DFEArrayType < DFEVar > arrayType =
    new DFEArrayType < DFEVar >(floatType, n);
```

Coefficients W_N^k are not calculated on the DFE; instead, they are calculated during bitstream compilation and are used as hardware constants.

Fig. 3.4 Graphical illustration of the manager of the Cooley-Tukey radix 2 algorithm

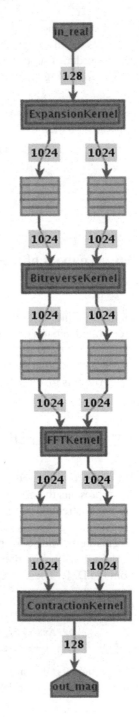

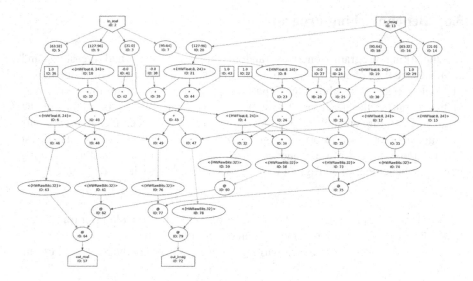

Fig. 3.5 Graphical illustration of kernel doing the fast Fourier transform calculation. This figure was generated using maxRenderGraphs command. Command maxRenderGraphs is part of the software package that is delivered with Maxeler compiler [Maxeler2015]. The illustration represents DataFlow graph for calculating the parallel radix 2 Cooley-Tukey algorithm for calculating the fast Fourier transform of the input sequence with four input points

3.7.1 Using DFE for Calculating the Fast Fourier Transform from C\C++ Code

MaxCompiler generates header file **FFT.h** that contains the following method:

```
FFT (
            int        number_of_input_sequences,
            float *imag_in,
            float *real_in,
            float *imag_out,
            float *real_out
    );
```

Calling this method from C\C++ code will send data from the CPU to the DFE, do calculation of fast Fourier transform on the DFE, and send results back to the CPU. For MaxCompiler to generate header file FFT.h, SLiC option needs to be turned on.

3.8 Benchmarking Program

Benchmarking program generates random input sequences and measures and compares the time of execution for both the original and accelerated implementation of Cooley-Tukey algorithm. Benchmarking program also checks the correctness of the output of the accelerated algorithm by comparing it to the output of the original CPU implementation for the same input sequence. If the user wants benchmarking program to print out generated input sequences and calculated fast Fourier transforms of those sequences, the user should uncomment the following line in the benchmarking program C++ code and recompile it:

```
#define PRINT_OUTPUT
```

Benchmarking program accepts a number of consecutive fast Fourier transforms that will be calculated as a parameter from the command line.

Entire C++ code of the benchmarking program is given in the Sect. 3.12 of this book.

3.9 Performance Measurements of the Accelerated Algorithm and Analysis of the Results

3.9.1 Assumptions and Limitations

Results of all the measurements shown in this chapter are measured on MAX3 system (MAX3424A card) using only one DFE. On this system we were able to synthesize the hardware for calculating the fast Fourier transform of the input sequence that has the maximum length of 32 input points.

All results of the fast Fourier transform algorithm performance on multiprocessor systems, shown in this chapter, are measured using only one processor core.

3.9.2 Types of Experiments

In our experiments we varied the size of the input sequence and the number of consecutive calculations of the fast Fourier transform. We measured the execution time of the original and accelerated algorithm for 100, 1,000, 10,000, 100,000, 1,000,000, 10,000,000 consecutive calculations of the fast Fourier transform for the input sequences with lengths of 8, 16, and 32 points.

Based on the results of experiments, the following graphs have been generated:

- Graphs that show performance of accelerated implementation compared with other publicly available implementations of various fast Fourier transform algorithms and implementations
- Graphs that show dependency between the total time needed to calculate consecutive fast Fourier transforms and the total number of consecutive transformations calculated
- Graphs that show achieved acceleration, speedup compared to original code, for the input sequence of lengths 8, 16, and 32 points.

Acceleration is calculated using the following formula:

$$\eta = \frac{T_{cpu}}{T_{max}}$$

T_{cpu} represents time it takes for an experiment to complete on CPU, and T_{max} represents time it takes for the same experiment to complete on Maxeler DFE.

3.9.3 Comparison of the Performance of the Accelerated Algorithm with Other Publicly Available Implementations

In this experiment we compared average run time needed to calculate 10,000,000 consecutive fast Fourier transform of accelerated algorithm with average time of best publicly available implementations of the fast Fourier transform algorithms running on the following machines:

- IBM QS20 Cell Blade
- PlayStation 3
- 3.60 GHz Intel Xeon Pentium 4 (Prescott)

The results of the experiments are shown in following three graphs (Graphs 3.1, 3.2, and 3.3). Graphs are showing the 30 fastest implementations and their average time of calculating the fast Fourier transform for the input sequence of lengths 8, 16, and 32 points.

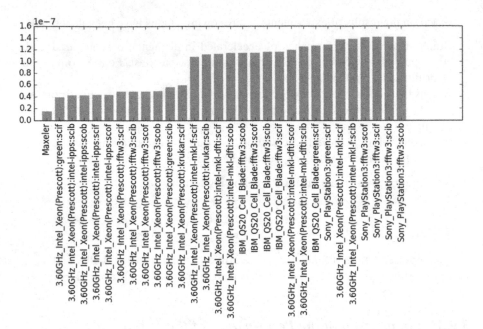

Graph 3.1 Average execution time in seconds of the best publicly available algorithms for calculating the fast Fourier transform on different computer architectures for the input sequence of eight points. The graph shows that implementation accelerated using Maxeler DFE has the shortest average execution time when compared with other publicly available implementations

Labels on y-axis are in the following format:

<details about the machine that test was run on>:<algorithm/implementation>:sc(i|o)(f|b)

s – algorithm works in single precision

c – algorithm is performing complex fast Fourier transform

i – algorithm works in-place

o – algorithm does not work in-place

f – forward transformation

b – backward transformation

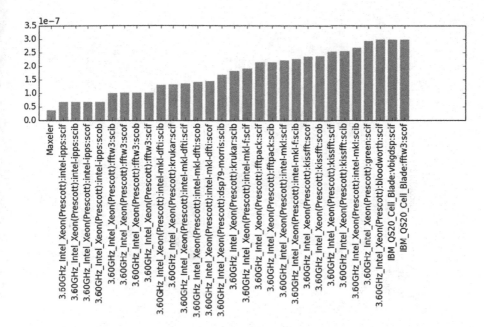

Graph 3.2 Average execution time in seconds of the best publicly available algorithms for calculating the fast Fourier transform on different computer architectures for the input sequence of 16 points. The graph shows that implementation accelerated using Maxeler DFE has the shortest average execution time when compared with other publicly available implementations

Labels on y-axis are in the following format:

<details about the machine that test was run on>:<algorithm/implementation>:sc(i|o)(f|b)

s – algorithm works in single precision

c – algorithm is performing complex fast Fourier transform

i – algorithm works in-place

o – algorithm does not work in-place

f – forward transformation

b – backward transformation

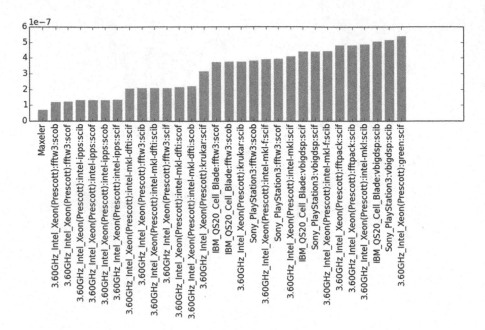

Graph 3.3 Average execution time in seconds of the best publicly available algorithms for calculating the fast Fourier transform on different computer architectures for the input sequence of thirty-two points. The graph shows that implementation accelerated using Maxeler DFE has the shortest average execution time when compared with other publicly available implementations
Labels on y-axis are in the following format:
<details about the machine that test was run on>:<algorithm/implementation>:sc(i|o)(f|b)
s – algorithm works in single precision
c – algorithm is performing complex fast Fourier transform
i – algorithm works in-place
o – algorithm does not work in-place
f – forward transformation
b – backward transformation

3.9.3.1   Data Preparation

The bash script for filtering files with the raw data about average execution speed of publicly available implementations of the fast Fourier transform on different machines is given in the Sect. 3.12 of this book. The script filters and converts the data to the format that will be used by the python script for drawing graphs. Files with the raw data are generated by Matteo Frigo and Steven G. Johnson [Frigo2014] as part of their research.

The bash, data preparation, script needs to be executed from the directory containing .speed files. As the result of the script execution for every .speed file new .filtered.speed file will be generated, also the file named all.filtered.speed will be generated.

3.9.4 Comparison of the Performance of Accelerated Algorithm with Original Implementation

The next seven graphs (Graphs 3.4, 3.5, 3.6, 3.7, 3.8, 3.9, and 3.10) show the comparison of performance of the accelerated algorithm compared to the original CPU implementation. All graphs that show execution time are given in two forms. The first form shows the execution time on a linear scale, while the second form shows execution time on a logarithmic scale. The second form is more suitable for observing some phenomena.

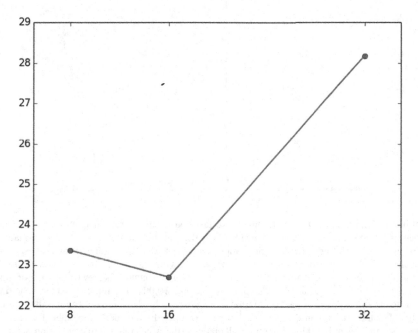

Graph 3.4 Dependency of achieved acceleration from the number of points in the input sequence. The graph shows acceleration of the original CPU Cooley-Tukey algorithm implementation achieved using the accelerated Maxeler DFE implementation. The experiment used to generate data consisted of 10,000,000 consecutive calculations of the fast Fourier transform on randomly generated input sequences. From the graph we can observe that the biggest acceleration is achieved when the length of the input sequence is 32 points; in that case the accelerated algorithms achieve acceleration of 28 times

3.9.4.1   Results of Experiments for Input Sequence of Eight Points

Execution time

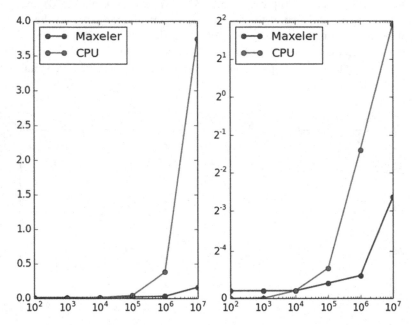

Graph 3.5 Total execution time of consecutive fast Fourier transform calculations for the input sequences of length 8 in seconds depending on the number of consecutive transformations performed for the accelerated and original implementations. The *left* graph is using a linear scale, while the *right* graph is using a logarithmic scale for showing execution time on *y*-axis. From the *right* graph, it could be observed that the accelerated implementation is slower than the original one if the number of consecutive calculations is less than 10,000. The reason for this is the time needed to configure Maxeler card and initial latency while the data starts streaming through the pipeline. For the 10,000 consecutive calculations, execution time of both implementations is the same. For more than 10,000 consecutive calculations, the accelerated implementation achieves faster execution time than the original implementation. The conclusion of this observation is that the accelerated version, for the case of the eight-point input sequence, should only be used if we want to perform more than 10,000 consecutive calculations

Acceleration

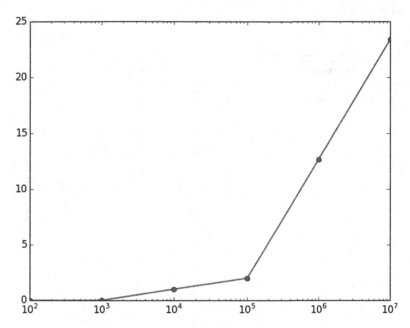

Graph 3.6 Graph of achieved acceleration and of the accelerated implementation of the Cooley-Tukey algorithm compared to the original implementation, depending on the number of consecutive fast Fourier transformation calculations for the input sequences of 8 points. From the graph, we could observe that with the increase of the number of consecutive fast Fourier transform calculations performed, there is increase in the achieved acceleration. Acceleration is smaller than 1 for less than 10,000 consecutive calculations. For the 10,000 calculations, the acceleration is 1, while for more than 10,000 consecutive calculation, the acceleration is greater than 1

3.9.4.2   Results of Experiments for Input Sequence of 16 Points

Execution time

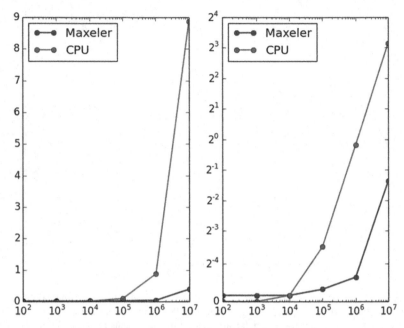

Graph 3.7 Total execution time of consecutive fast Fourier transform calculations for the input sequences of length 16 in seconds depending on the number of consecutive transformations performed for the accelerated and original implementations. The *left* graph is using a linear scale, while the *right* graph is using a logarithmic scale for showing execution time on *y*-axis. From the *right* graph, it could be observed that the accelerated implementation is slower than the original one if the number of consecutive calculations is less than 10,000. The reason for this is the time needed to configure Maxeler card and initial latency while the data starts streaming through the pipeline. For the 10,000 consecutive calculations, the execution time of both implementations is the same. For more than 10,000 consecutive calculations, the accelerated implementation achieves faster execution time than the original implementation. The conclusion of this observation is that the accelerated version, for the case of the eight-point input sequence, should only be used if we want to perform more than 10,000 consecutive calculations

Acceleration

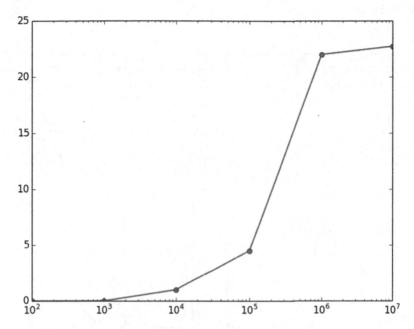

Graph 3.8 Graph of achieved acceleration and of the accelerated implementation of the Cooley-Tukey algorithm compared to the original implementation, depending on the number of consecutive fast Fourier transformation calculations for the input sequences of 16 points. From the graph, we could observe that with the increase of the number of consecutive fast Fourier transform calculations performed, there is increase in the achieved acceleration. Acceleration is smaller than 1 for less than 10,000 consecutive calculations. For the 10,000 calculations, the acceleration is 1, while for more than 10,000 consecutive calculation, the acceleration is greater than 1

3.9.4.3   Results of Experiments for Input Sequence of 32 Points

Execution time

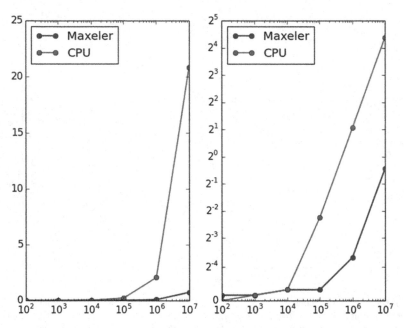

Graph 3.9 Total execution time of consecutive fast Fourier transform calculations for the input sequences of length 32 in seconds depending on the number of consecutive transformations performed for the accelerated and original implementations. The *left* graph is using a linear scale, while the *right* graph is using a logarithmic scale for showing execution time on y-axis. From the *right* graph, it could be observed that the accelerated implementation is slower than the original one if the number of consecutive calculations is less than 1,000. The reason for this is the time needed to configure Maxeler card and initial latency while the data starts streaming through the pipeline. For the 1,000–10,000 consecutive calculations, execution time of both implementations is the same. For more than 10,000 consecutive calculations, the accelerated implementation achieves faster execution time than the original implementation. The conclusion of this observation is that the accelerated version, for the case of the eight-point input sequence, should only be used if we want to perform more than 10,000 consecutive calculations

Acceleration

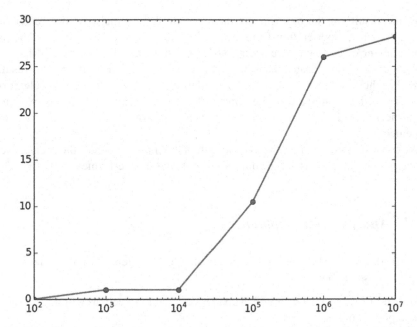

Graph 3.10 Graph of achieved acceleration and of the accelerated implementation of the Cooley-Tukey algorithm compared to the original implementation, depending on the number of consecutive fast Fourier transformation calculations for the input sequences of 32 points. From the graph, we could observe that with the increase of the number of consecutive fast Fourier transform calculations performed, there is increase in the achieved acceleration. Acceleration is smaller than 1 for less than 1,000 consecutive calculations. For the 1,000–10,000 calculations, the acceleration is 1, while for more than 10,000 consecutive calculations, the acceleration is greater than 1

3.9.5   Graph Drawing Script

The script for drawing graphs shown in this chapter is written in Python 2.7. In order to execute the script, Python package matplotlib [Hunter2012] needs to be installed. Matplotlib is a free package that could be installed with (e.g., on OSX) *pip install matplotlib* command. The package allows drawing of graphs in a similar way as if using MATLAB. The results of the experiments needed to generate graphs are embedded into the source code of the graph drawing Python script.

The entire graph drawing script is given in the Sect. 3.12 of this book.

3.9.6 Results Overview and Performance Predictions

From the results of the experiments that measured the time needed to calculate consecutive fast Fourier transformations, we can come to the conclusion that some number of consecutive calculations is needed for the accelerated, Maxeler, implementation to be faster than the original, CPU, implementation. The results show that the number of consecutive calculations needed to achieve acceleration decreases with the increase of the length of the input sequence. This is expected as the accelerated implementation has lower time complexity than the original implementations.

We can also notice that for the fixed length of the input sequence, the acceleration increases with the increase of the number of calculations performed.

3.9.7 Analysis of the Bottlenecks

The performance predictions are not true if there is a bottleneck in the system.
   List of possible bottlenecks:

- Transfer of data to and from the Maxeler card cannot keep up with the speed of data processing on the card.
  - In this case bottleneck is in the I/O controller.
  - Amount of data that should be transferred to and from the Maxeler card increases linearly with the length of the input sequence.
- The Maxeler card does not have enough hardware resources to synthesize the Cooley-Tukey radix 2 algorithm.
  - In this case bottleneck is in the number of hardware resources available on the Maxeler card.
  - The number of needed hardware resources increases with dependence $O(N \log N)$, where N is the length of the input sequence.

Explanation of the Hardware Resource Usage of the Cooley-Tukey Radix 2 Algorithms: Let N be the number of points in input sequence. The radix 2 algorithm will have $\log N$ phases. Each phase needs $O(N)$ hardware resources to be implemented. This yields that the total number of hardware resources needed to implement the radix 2 algorithm is $O(N)$.

3.10 Possible Modifications and Improvements

Initial latency of calculation can be reduced by using the radix r algorithm with r bigger than 2. The downside of this approach is more complicated bitstream using more of the limited hardware resources. Often the initial latency is of no significant importance; instead, the goal is to maximize the throughput of the system, in our case the number of fast Fourier transform calculations in one unit of time, and to minimize the resource utilization.

Because of the limited hardware resources on one Maxeler card, there is the limitation on the input sequence length for which hardware implementation of radix 2 algorithm could be synthesized. This limitation could be overcome by: i) splitting the fast Fourier transform calculation into multiple phases where the output of one phase is the input of the next phase and ii) implementing these phases as different kernel. This approach could be used for calculating the fast Fourier transform of larger input sequences using more than one DFE (Maxeler card).

3.11 Conclusion

The performed experiments have shown that the accelerated implementation has the expected performance and that it produces equivalent results to the original implementation. The performance of accelerated application has surpassed the performance of other publicly available implementation of the fast Fourier transform. Because of the limited number of hardware resource on the Maxeler card used for our experiments, we were not able to generate a bitstream for calculating fast Fourier transform of input sequences with more than 32 points. On better Maxeler machines, like MAX4 (Maia or Coria), it could be possible to generate a bitstream that calculates the fast Fourier transform of longer input sequences.

Because of the properties of radix 2 algorithm, it is possible to make a hybrid solution that would use both CPU and Maxeler DFE for calculating fast Fourier transform. The CPU part of the application would divide the problem into smaller problems that would, depending on their size, either be solved on the DFE or divided again.

It is expected, and performed experiments confirm, that there is increase in achieved acceleration with the increase in input sequence number of elements. Acceleration is increasing because the initial latency of the DFE implementation is increasing slower than the latency of the CPU implementation. Also when performing multiple calculations, after the period of initial latency, the DFE implementation produces results on every clock cycle, while the CPU implementation has the same latency for calculating every result.

To achieve speedup using the accelerated implementation, it is necessary to perform multiple consecutive calculations of the fast Fourier transform. The number of consecutive calculations needed to perform in order to achieve acceleration compared to the original CPU implementation depends on the number of points in the input sequence. The method for determining this number is shown in section "Comparison of the Performance of Accelerated Algorithm with Original Implementation.

3.12   Original Implementation Code

3.12.1   *MaxJ Code of Kernel that Calculates Fast Fourier Transform*

```
import com.maxeler.maxcompiler.v2.kernelcompiler.Kernel;
import com.maxeler.maxcompiler.v2.kernelcompiler.KernelParameters;
import com.maxeler.maxcompiler.v2.kernelcompiler.types.base.DFEType;
import com.maxeler.maxcompiler.v2.kernelcompiler.types.base.DFEVar;
import com.maxeler.maxcompiler.v2.kernelcompiler.types.composite.DFEArray;
import com.maxeler.maxcompiler.v2.kernelcompiler.types.composite.DFEArrayType;

class FFTKernel extends Kernel {

    private static final DFEType floatType = dfeFloat(8, 24);

    protected FFTKernel(KernelParameters parameters, int n) {
        super(parameters);

        DFEArrayType < DFEVar > arrayType =
            new DFEArrayType < DFEVar > (floatType, n);

        DFEArray < DFEVar > in_real = io.input("in_real", arrayType);
        DFEArray < DFEVar > in_imag = io.input("in_imag", arrayType);


        int log_n = Integer.numberOfTrailingZeros(n);

        DFEVar[][] real = new DFEVar[log_n + 1][n];
        DFEVar[][] imag = new DFEVar[log_n + 1][n];

        for (int i = 0; i < n; i++) {
            int j = Integer.reverse(i) >>> (32 - log_n);

            real[0][i] = in_real[j];
            imag[0][i] = in_imag[j];
        }

        for (int i = 1, log_i = 0; i < n; i *= 2, log_i++) {
            for (int k = 0; k < i; k++) {
                float w_real = (float) Math.cos(-2 * Math.PI * k / (i * 2));
                float w_imag = (float) Math.sin(-2 * Math.PI * k / (i * 2));

                for (int j = 0; j < n; j += i * 2) {
                    DFEVar temp_real =
                        real[log_i][j + k + i] * w_real - imag[log_i][j + k + i] * w_imag;
```

```
                    DFEVar temp_imag =
                        real[log_i][j + k + i] * w_imag + imag[log_i][j + k + i] * w_real;

                    real[log_i+ 1][j + k + i] = real[log_i][j + k] - temp_real;
                    imag[log_i+ 1][j + k + i] = imag[log_i][j + k] - temp_imag;

                    real[log_i+ 1][j + k] = real[log_i][j + k] + temp_real;
                    imag[log_i+ 1][j + k] = imag[log_i][j + k] + temp_imag;
                }
            }
        }

        DFEArray < DFEVar > out_real = arrayType.newInstance(this);

        for (int i = 0; i < n; i++)
            out_real[i] <= = real[log_n][i];

        io.output("out_real", out_real, arrayType);

        DFEArray < DFEVar > out_imag = arrayType.newInstance(this);

        for (int i = 0; i < n; i++)
            out_imag[i] <== imag[log_n][i];

        io.output("out_imag", out_imag, arrayType);

        }
    }
```

3.12.2   MaxJ Manager Code of Cooley-Tukey Radix 2 Algorithm

```
package fft;

import com.maxeler.maxcompiler.v2.build.EngineParameters;
import com.maxeler.maxcompiler.v2.managers.custom.CustomManager;
import com.maxeler.maxcompiler.v2.managers.custom.DFELink;
import com.maxeler.maxcompiler.v2.managers.custom.blocks.KernelBlock;
import com.maxeler.maxcompiler.v2.managers.custom.stdlib.Max3RingConnection;
import com.maxeler.maxcompiler.v2.managers.custom.stdlib.MaxRingBidirectionalStream;
import com.maxeler.maxcompiler.v2.managers.engine_interfaces.CPUTypes;
import com.maxeler.maxcompiler.v2.managers.engine_interfaces.EngineInterface;
import com.maxeler.maxcompiler.v2.managers.engine_interfaces.InterfaceParam;

public class FFTManager extends CustomManager {
```

```
public FFTManager(EngineParameters engineParameters, int n) {
super(engineParameters);
KernelBlock bitreversePhase = addKernel(
new BitReverseKernel(makeKernelParameters("BitreversePhase"), n));

KernelBlock fftPhase = addKernel(
new FFTKernel(makeKernelParameters("FFTPhase"), n, 0, Integer.numberOfTrailingZeros(n)/2, false, true));

DFELink in_real = addStreamFromCPU("in_real");
DFELink in_imag = addStreamFromCPU("in_imag");

bitreversePhase.getInput("in_real") <== in_real;
bitreversePhase.getInput("in_imag") <== in_imag;

fftPhase.getInput("in_real") <== bitreversePhase.getOutput("out_real");
fftPhase.getInput("in_imag") <== bitreversePhase.getOutput("out_imag");

MaxRingBidirectionalStream real_ring = addMaxRingBidirectionalStream("realRingStream",
Max3RingConnection.MAXRING_A);
MaxRingBidirectionalStream imag_ring = addMaxRingBidirectionalStream("imagRingStream",
Max3RingConnection.MAXRING_A);

fftPhase.getInput("in_real_dummy") <== real_ring.getLinkFromRemoteDFE();
fftPhase.getInput("in_imag_dummy") <== imag_ring.getLinkFromRemoteDFE();

real_ring.getLinkToRemoteDFE() <== fftPhase.getOutput("out_real");
imag_ring.getLinkToRemoteDFE() <== fftPhase.getOutput("out_imag");
}

static EngineInterface interfaceDefault (int n) {
EngineInterface ei = new EngineInterface();
InterfaceParam size = ei.addParam("dataSize", CPUTypes.UINT64);

ei.setTicks("FFTPhase", size);
ei.setTicks("BitreversePhase", size);

ei.setStream("in_real", CPUTypes.FLOAT, size * n * CPUTypes.FLOAT.sizeInBytes());
ei.setStream("in_imag", CPUTypes.FLOAT, size * n * CPUTypes.FLOAT.sizeInBytes());

return ei ;
}

public static void main(String[] args) {
EngineParameters params = new EngineParameters(args);
```

```
          int n = Integer.parseInt(System.getenv("n"));

          System.out.println(n);

          FFTManager manager = new FFTManager(params, n);

          //manager.setIO(IOType.ALL_CPU);
          manager.addMaxFileConstant("n", n);
          manager.createSLiCinterface(interfaceDefault(n));

          //manager.getBuildConfig().setBuildEffort(Effort.LOW);
          manager.getBuildConfig().setMPPRCostTableSearchRange(1, 3);
          manager.getBuildConfig().setMPPRParallelism(3);

          manager.build();
          }
          }
```

3.12.3   MaxJ Manager Code of Bit Reverse Kernel

```
package fft;

import com.maxeler.maxcompiler.v2.kernelcompiler.Kernel;
import com.maxeler.maxcompiler.v2.kernelcompiler.KernelParameters;
import com.maxeler.maxcompiler.v2.kernelcompiler.types.base.DFEType;
import com.maxeler.maxcompiler.v2.kernelcompiler.types.base.DFEVar;
import com.maxeler.maxcompiler.v2.kernelcompiler.types.composite.DFEArray;
import com.maxeler.maxcompiler.v2.kernelcompiler.types.composite.DFEArrayType;

class BitReverseKernel extends Kernel {

private static final DFEType floatType = dfeFloat(8,24);

protected BitReverseKernel(KernelParameters parameters, int n) {
super(parameters);

DFEArrayType<DFEVar> arrayType = new DFEArrayType<DFEVar>(floatType, n);

DFEArray<DFEVar> in_real = io.input("in_real", arrayType);
DFEArray<DFEVar> in_imag = io.input("in_imag", arrayType);


int log_n = Integer.numberOfTrailingZeros(n);

DFEVar[] real = new DFEVar[n];
DFEVar[] imag = new DFEVar[n];
```

```java
for(int i = 0; i < n; i++) {
int j = Integer.reverse(i) >>> (32 - log_n);

 real[i] = in_real[j];
 imag[i] = in_imag[j];
}


DFEArray<DFEVar> out_real = arrayType.newInstance(this);

for (int i = 0; i < n; i++)
out_real[i] <== real[i];


io.output("out_real", out_real, arrayType);


DFEArray<DFEVar> out_imag = arrayType.newInstance(this);

for (int i = 0; i < n; i++)
out_imag[i] <== imag[i];


io.output("out_imag", out_imag, arrayType);


}


}
```

3.12.4 C++ Code of Radix 2 In-Place Cooley-Tukey Algorithm

```cpp
#include <cstdio>
#include<cmath>
#include<algorithm>

using namespace std;

#define PI 3.14159265359

// Reversing the bits in the number
unsigned bit_reverse(register unsigned x) {
  x = (((x & 0xaaaaaaaa) >> 1) | ((x & 0x55555555) << 1));
  x = (((x & 0xcccccccc) >> 2) | ((x & 0x33333333) << 2));
  x = (((x & 0xf0f0f0f0) >> 4) | ((x & 0x0f0f0f0f) << 4));
  x = (((x & 0xff00ff00) >> 8) | ((x & 0x00ff00ff) << 8));

  return ((x >> 16) | (x << 16));
}
```

```
// Cooley-Tukey radix-2 in-place  fast Fourier transfrom algorithm
void fft(float real[], float imag[], unsigned n) {
  // Reordering of array to bit-reverse order
  unsigned log_n = __builtin_ctz(n);

  for(unsigned i = 0; i < n; i++) {
    unsigned j = bit_reverse(i) >> ( sizeof(i)*8 - log_n);

    if(i > j) {
      swap(real[i], real[j]);
      swap(imag[i], imag[j]);
    }
  }

  for(unsigned i = 1; i < n; i <<= 1) {
    for(unsigned k = 0; k < i; k++) {
      float w_real = cos(-2*PI*k/(i<<1));
      float w_imag = sin(-2*PI*k/(i<<1));

      for(unsigned j = 0; j < n; j += i<<1) {
        float temp_real = real[j+k+i]*w_real - imag[j+k+i]*w_imag;
        float temp_imag = real[j+k+i]*w_imag + imag[j+k+i]*w_real;

        real[j+k+i] = real[j+k] - temp_real;
        imag[j+k+i] = imag[j+k] - temp_imag;

        real[j+k] += temp_real;
        imag[j+k] += temp_imag;
      }
    }
  }
}
```

3.12.5   C++ Benchmark Code

```cpp
#include "Maxfiles.h"
#include "MaxSLiCInterface.h"

#include<cstdio>
#include<cstdlib>
#include<ctime>

// #define PRINT_OUTPUT

const unsigned max_number_of_consecutive_calculations = 10000000;

int main(int argcnt, char* args[]) {

unsigned number_of_consecutive_calculations = argcnt == 2 ? atoll(args[1]) : 1;

  if(number_of_consecutive_calculations > max_number_of_consecutive_calculations) {
      printf("Maximum number of consecutive calculations: %u\n",
max_number_of_consecutive_calculations);

      number_of_consecutive_calculations = max_number_of_consecutive_calculations;
  }

  printf("Number of calculations:: %u\nInput sequence length:: %u\n",
         number_of_consecutive_calculations,
         FFT_n);

  static float real[number_of_consecutive_calculations][FFT_n];
  static float imag[number_of_consecutive_calculations][FFT_n];

  for(unsigned i = 0; i < number_of_consecutive_calculations ; i++)
    for(unsigned j = 0; j < FFT_n; j++) {
      real[i][j] = rand()%10;
      imag[i][j] = rand()%10;
    }

  printf("Generation of random input sequences completed\n");

#ifdef PRINT_OUTPUT
    for(unsigned k = 0; k < number_of_consecutive_calculations; k++)
    for(unsigned i = 0; i < FFT_n; i++)
      printf("%.3f + %.3fj\n", real[k][i], imag[k][i]);
#endif
```

```
static float o_real[number_of_consecutive_calculations][FFT_n];
static float o_imag[number_of_consecutive_calculations][FFT_n];

clock_t startTime;

printf("DFE:\n");

startTime = clock();

FFT(number_of_consecutive_calculations, (float *)imag, (float *)real, (float *)o_imag, (float *)o_real);

printf("Execution time: %f Normalised time: %e\n",

          ((double)(clock()-startTime))/CLOCKS_PER_SEC,
          ((double)(clock()-startTime))/CLOCKS_PER_SEC/number_of_consecutive_calculations);

#ifdef PRINT_OUTPUT
  for(unsigned k = 0; k < number_of_consecutive_calculations; k++)
   for(unsigned i = 0; i < FFT_n; i++)
     printf("%.3f + %.3fj\n", o_real[k][i], o_imag[k][i]);
#endif

  printf("CPU:\n");

  startTime = clock();
  for(unsigned k = 0; k < number_of_consecutive_calculations; k++) {
  fft(real[k], imag[k], FFT_n);
  }

  printf("Execution time: %f Normalised time: %e\n",
         ((double)(clock()-startTime))/CLOCKS_PER_SEC,
         ((double)(clock()-startTime))/CLOCKS_PER_SEC/number_of_consecutive_calculations);

#ifdef PRINT_OUTPUT
  for(unsigned k = 0; k < number_of_consecutive_calculations; k++)
   for(unsigned i = 0; i < FFT_n; i++)
     printf("%.3f + %.3fj\n", real[k][i], imag[k][i]);
#endif
  for(unsigned k = 0; k < number_of_consecutive_calculations; k++)
   for(unsigned i = 0; i < FFT_n; i++)
    if(real[k][i] - o_real[k][i] > 0.1e-10 || imag[k][i] - o_imag[k][i] > 0.1e-10) {
       return 1;
     }

  printf("All done");
  return 0;
}
```

3.12.6   Script for Filtering Raw Data About Average Execution Speed of the Experiments

#!/bin/bash

```
grep -E 'sc.. ((8)|(16)|(32)) "find . -name "*.speed" ! -name
"*filtered.speed" -maxdepth 1 -print'| awk '{
sub(/.speed/, ""); print $1, $2, $3, $5}'
> 'all.filtered.speed'

for i in 'find . -name "*.speed" ! -name "*filtered.speed"
    -maxdepth 1 -print'; do
    grep -E 'sc.. ((8)|(16)|(32)) '"$i" | awk '{print $1, $2,
    $3, $5}'  > 'echo "$i" | sed 's/speed/filtered.speed/''
done
```

3.12.7   Python Script for Automatic Graph Generation

```python
from pylab import *

scale = [100,    1000,    10000,    100000,    1000000,    10000000];

# Experiment results
max_32 = [0.010000, 0.010000, 0.020000, 0.020000, 0.080000, 0.740000];
cpu_32 = [0.000000, 0.010000, 0.020000, 0.210000, 2.080000, 20.850000];

max_16 = [0.010000, 0.010000, 0.010000, 0.020000, 0.040000, 0.390000];
cpu_16 = [0.000000, 0.000000, 0.010000, 0.090000, 0.880000, 8.860000];

max_8 = [0.010000, 0.010000, 0.010000, 0.02000, 0.030000, 0.160000];
cpu_8 = [0.000000, 0.000000, 0.010000, 0.040000, 0.380000, 3.740000];

labels = ["Maxeler", "CPU"];

# Draw execution time and acceleration for experiment
def draw_experiment(input_len, scale, labels, times):
    acceleration = [t_cpu / t_max for t_max, t_cpu in zip(times[0], times[1])]

    plot(scale, acceleration, "-o" , color="red", linewidth=1.5)
    xscale('log');

    savefig('acceleration.'+ str(input_len) + ".png");
    clf()
    for time, label in zip(times, labels):
        plot(scale, time, "-o", label=label, linewidth=1.5)

    legend(loc='upper left')
```

```
    xscale('log');
    savefig('time.'+ str(input_len) + ".png");
    yscale('symlog', basey = 2, linthreshy = 1e-1);
    savefig('time.'+ str(input_len) + ".log.png");
    clf()

import sys
from collections import defaultdict

infile = 'all.filtered.speed';

data = {8 : [( "Maxeler", max_8[-1]/scale[-1])], 16 : [( "Maxeler", max_16[-1]/scale[-1])], 32 : [( "Maxeler",
max_32[-1]/scale[-1])]}

with open(infile) as inf:
    for line in inf:
        line_words = line.split()
        data[int(line_words[2])].append((line_words[0].strip("./") + ":" + line_words[1], float(line_words[3])))

import heapq

# Draw side by side comparison graph
for size in (8, 16, 32):
    data[size] = heapq.nsmallest(20, data[size], key=lambda d:d[1])

    labels, speeds = zip(*data[size])
    gcf().set_size_inches(12, 8)

    from mpl_toolkits.axes_grid1.axes_divider import make_axes_area_auto_adjustable

    make_axes_area_auto_adjustable(gca())

    bar(range(1, len(speeds) + 1),
            speeds, facecolor='#9999ff',
            edgecolor='white',
            align='center')
    xticks(range(1, len(speeds) + 1), labels, rotation='vertical'
    savefig('perf.'+ str(size) + ".log.png");
    clf()

draw_experiment(16, scale, labels, [max_16, cpu_16]);
draw_experiment(32, scale, labels, [max_32, cpu_32]);
draw_experiment(8, scale, labels, [max_8, cpu_8]);

# Draw acceleration/number of input points Graph
plot([8, 16, 32], [cpu_8[-1]/max_8[-1], cpu_16[-1]/max_16[-1], cpu_32[-1]/max_32[-1]], "-o", color="red",
linewidth=1.5)
xticks([8, 16, 32])

savefig('acceleration.input_length.png');
clf()
```

References

[Cooley1969] Cooley J, Lewis P, Welch P (1969) The fast fourier transform and its applications. IEEE Trans Educ 12(1):77–85

[Frigo2014] (2014) FFT benchmark results [Online], December. Available: http://www.fftw.org/speed

[Hunter2012] Hunter J (2014) Matplotlib [Online], December. Available: http://matplotlib.org

[Maxeler2015] Maxeler (2015) Multiscale dataFlow programming. Maxeler Technologies Ltd, London

Chapter 4
Using the WebIDE

4.1 The WebIDE System Overview

This chapter utilizes the methodology that compares one figure with one thousand words. Therefore, the material to follow is presented via figures, graphs, pseudo code, and real code in a way which is without redundancy, but is perfectly understood by programmers (Fig. 4.1–4.17).

4.2 Installation Instructions

4.2.1 WebIDE Dependencies

To successfully install WebIDE, the following programs need to be installed first:

1. MaxCompiler
2. Python 2.7
3. Shellinabox

Python 2.7 and MaxCompiler should be preinstalled on MaxWorkstation. You might need to install Python development headers if you want to use WebIDE with Gunicorn web server.

To install Python development headers on CentOS, run the following command as a root user:

```
yum install python-devel
```

To install shellinabox on CentOS run following command as a root user:

```
yum install shellinabox
```

© Springer International Publishing Switzerland 2015
V. Milutinović et al., *Guide to DataFlow Supercomputing*, Computer
Communications and Networks, DOI 10.1007/978-3-319-16229-4_4

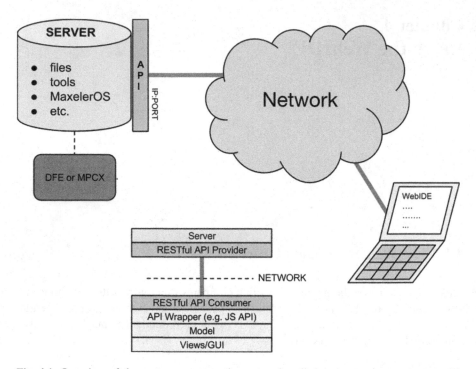

Fig. 4.1 Overview of the system structure; the system is split into two main components: (1) server component and (2) client component. Components communicate to each other using RESTful HTTP API (application programming interface) that server component exposes to the other components that want to interact with it

4.2.2 Installation Steps

```
# as a root
tar -zxvf maxapi-<version>.tar.gz
cd maxapi-<version>
python setup.py install
```

4.3 Running WebIDE

```
# as a root user
# This will start shellinabox and
# WebIDE test server  on port 5002
WebIDE.py
# This will start both shellinabox and
# Gunicorn server,
# serving WebIDE on port 5000.
startWebIDE
```

4.4 Security Notes

4.4.1 Note1

This version of WebIDE uses Unix for user authorization, but Unix security is not yet properly implemented (user programs currently run as root).

Additional Info Support for downgrading privileges to match the privileges of user that has sent request to the server is built in into WebIDE server, but is disabled because it causes thread that served the request to crash after the request was served.

4.4.2 Note2

WebIDE traffic should be encrypted if it is used over the Internet. NGINX could be used for this purpose (please see Deployment Notes for more details).

4.5 Deployment Notes

It's desired to use NGINX in front of Gunicorn and shellinabox for better performance. Using NGINX can provide additional benefits like load balancing, https support, etc (Fig. 4.2).

4.5.1 NGINX

NGINX is web server that can work as a reverse proxy. In (Fig. 4.2) NGINX is working as a reverse proxy server doing load balancing for the *WebIDE server* and for *shellinabox*. NGINX can also decrypt SSL sessions so he could accept HTTPS traffic and convert it to HTTP traffic.

4.5.2 Gunicorn

To best utilize available computing resource, it is often needed for a web server to run in multiple threads. Gunicorn is a WSGI HTTP server that allows us to run our WebIDE server using multiple threads.

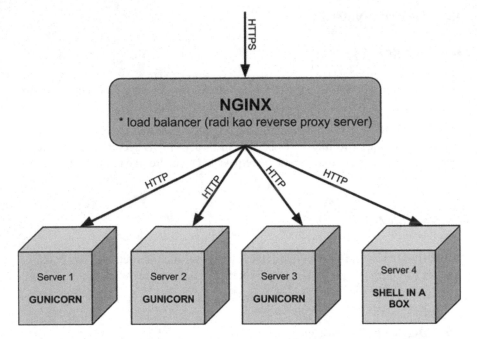

Fig. 4.2 Configuration of WebIDE server component; this picture presents desired way to configure and deploy WebIDE

4.6   User Administration

WebIDE currently uses Linux PAM (pluggable authentication module) for authentication. Depending on your system setup, authentication could be done using LDAP (Lightweight Directory Access Protocol) server.

Because WebIDE uses PAM for user authentication, any user that has access to the machine where WebIDE is installed would be able to access WebIDE with the same credentials it uses to access that machine.

On standard MaxWorkstation new user can be created using useradd command.

Note User credentials will, at some point, be unique across all Maxeler services including WebIDE, AppGallery, and MDX.

4.7   Login

Before doing any action on the system, the user needs to login first. Login process consists of user providing correct username and password. By providing correct username and password, the user proves that he has the right to access the system (Fig. 4.3).

Fig. 4.3 Screenshot of login view

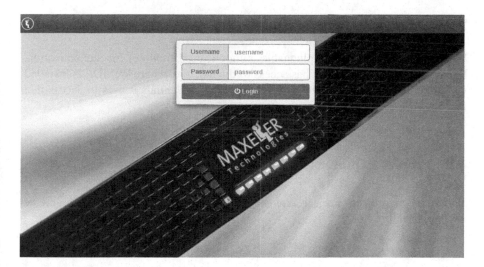

Fig. 4.4 Screenshot of login view after unsuccessful login

This view is used to authenticate user based on his Unix username and his password (Fig. 4.4).

This would happen when wrong credentials are entered.

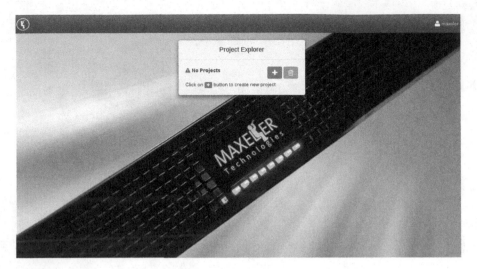

Fig. 4.5 Screenshot of Project Explorer view without any projects

4.7.1 Login Credentials

WebIDE uses Linux PAM (pluggable authentication module) for authentication. Because of this, any user that has access to the machine where WebIDE is installed would be able to access WebIDE with the same credentials it uses to access that machine.

4.8 Project Explorer (Fig. 4.5)

If there are no projects, Project Explorer view would show help about how to create new project (Fig. 4.6).

4.9 Creating a New Project

To create a new project, click on the ![+] button. After that you will be prompted with *Create new project* dialog that allows you to name your project and chose from which template project will be created. There is a large number of existing templates that you can choose to create a new project from. All tutorial examples and training exercises are listed as project templates. If you do not choose any template, your project will be created from template called *Default Template* (Fig. 4.7).

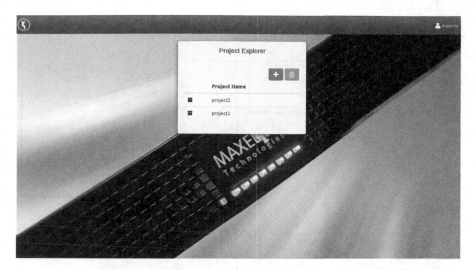

Fig. 4.6 Screenshot of Project Explorer view with two projects: project 1 and project 2

Fig. 4.7 Screenshot of Create new project dialog

4.10 Deleting an Existing Project

To delete an existing project:

(a) Click on it and it will become selected
    (you should see it has become highlighted as in picture 4.8-3).
(b) Click on the red button with bin image on it to delete selected project.

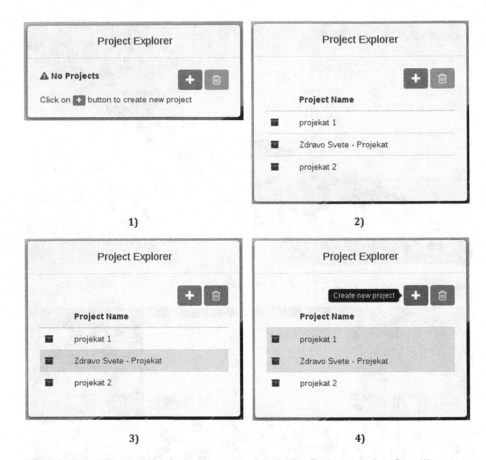

Fig. 4.8 Four different states of the components of the Project Explorer user interface. (1) presents the components in the case where the user has no projects. All others show the components in the event when the user has three existing projects (project 1, Hello World Project, Project 2). In (3) and (4) we can see that some of the projects are highlighted in *red*; that means that they are selected. The project is selected by clicking on its name. It is possible to select multiple projects. By pressing the *red* button with a picture of garbage, selected projects will be deleted. By pressing the *blue* button with a picture of the sign plus, it is possible to create a new project. (4) depicts the behavior that occurs when the user keeps longer mouse pointer over the button – the program displays a *black* note with white text that describes the function of the buttonA double click on the project's name opens the project (Color figure online)

It is possible to delete multiple projects at the same time. To achieve this keep the Shift button pressed on your keyboard while selecting multiple projects (Fig. 4.8).

4.11   WebIDE with Welcome Windows (Fig. 4.9)

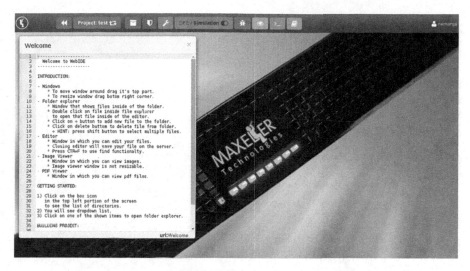

Fig. 4.9 Screenshot of IDE view with Welcome Windows

4.12   WebIDE with Side-by-Side Code Editors (Fig. 4.10)

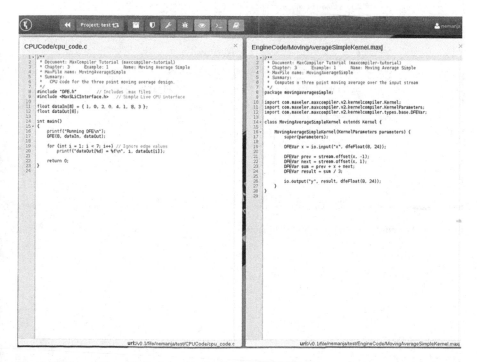

Fig. 4.10 Screenshot of IDE view with side-by-side code editors

4.13   Code Editor (Fig. 4.11)

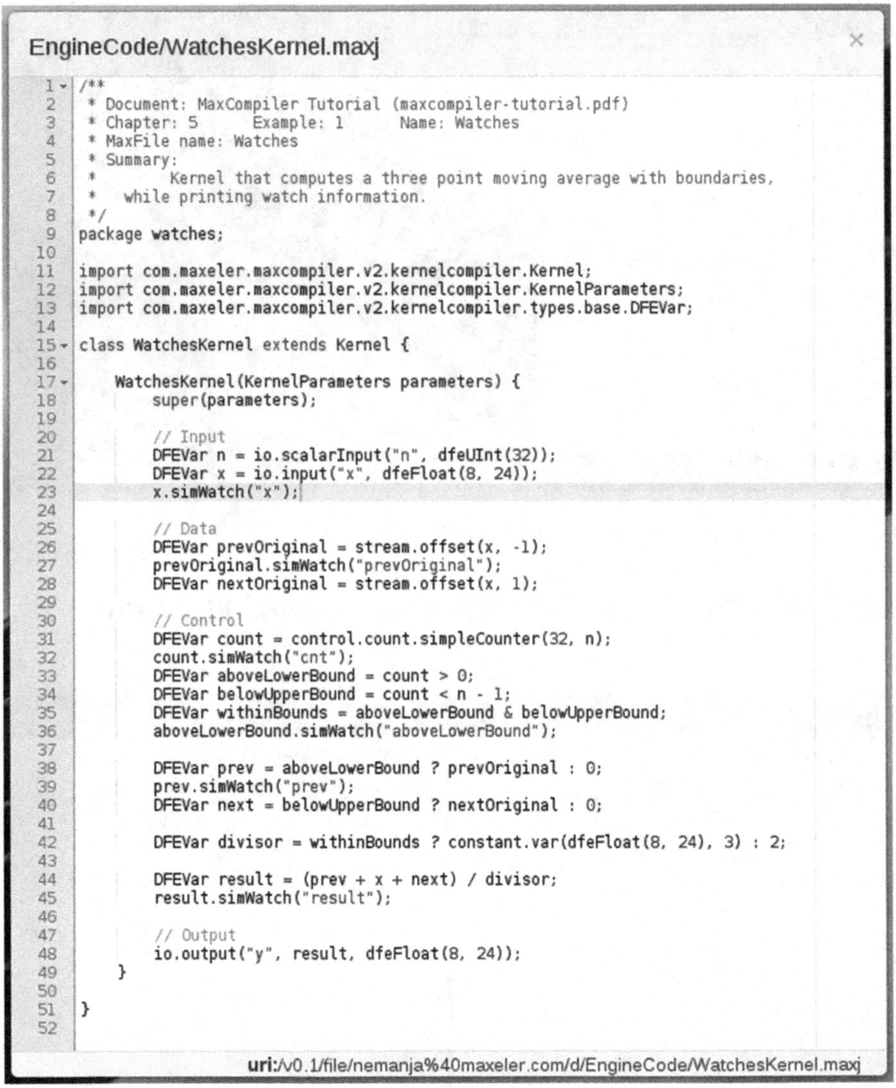

EngineCode/WatchesKernel.maxj   ✕

```
 1  /**
 2   * Document: MaxCompiler Tutorial (maxcompiler-tutorial.pdf)
 3   * Chapter: 5        Example: 1      Name: Watches
 4   * MaxFile name: Watches
 5   * Summary:
 6   *         Kernel that computes a three point moving average with boundaries,
 7   *   while printing watch information.
 8   */
 9  package watches;
10
11  import com.maxeler.maxcompiler.v2.kernelcompiler.Kernel;
12  import com.maxeler.maxcompiler.v2.kernelcompiler.KernelParameters;
13  import com.maxeler.maxcompiler.v2.kernelcompiler.types.base.DFEVar;
14
15  class WatchesKernel extends Kernel {
16
17      WatchesKernel(KernelParameters parameters) {
18          super(parameters);
19
20          // Input
21          DFEVar n = io.scalarInput("n", dfeUInt(32));
22          DFEVar x = io.input("x", dfeFloat(8, 24));
23          x.simWatch("x");
24
25          // Data
26          DFEVar prevOriginal = stream.offset(x, -1);
27          prevOriginal.simWatch("prevOriginal");
28          DFEVar nextOriginal = stream.offset(x, 1);
29
30          // Control
31          DFEVar count = control.count.simpleCounter(32, n);
32          count.simWatch("cnt");
33          DFEVar aboveLowerBound = count > 0;
34          DFEVar belowUpperBound = count < n - 1;
35          DFEVar withinBounds = aboveLowerBound & belowUpperBound;
36          aboveLowerBound.simWatch("aboveLowerBound");
37
38          DFEVar prev = aboveLowerBound ? prevOriginal : 0;
39          prev.simWatch("prev");
40          DFEVar next = belowUpperBound ? nextOriginal : 0;
41
42          DFEVar divisor = withinBounds ? constant.var(dfeFloat(8, 24), 3) : 2;
43
44          DFEVar result = (prev + x + next) / divisor;
45          result.simWatch("result");
46
47          // Output
48          io.output("y", result, dfeFloat(8, 24));
49      }
50
51  }
52
```

uri:/v0.1/file/nemanja%40maxeler.com/d/EngineCode/WatchesKernel.maxj

Fig. 4.11 Screenshot of Code Editor component

4.14 Build Output (Fig. 4.12)

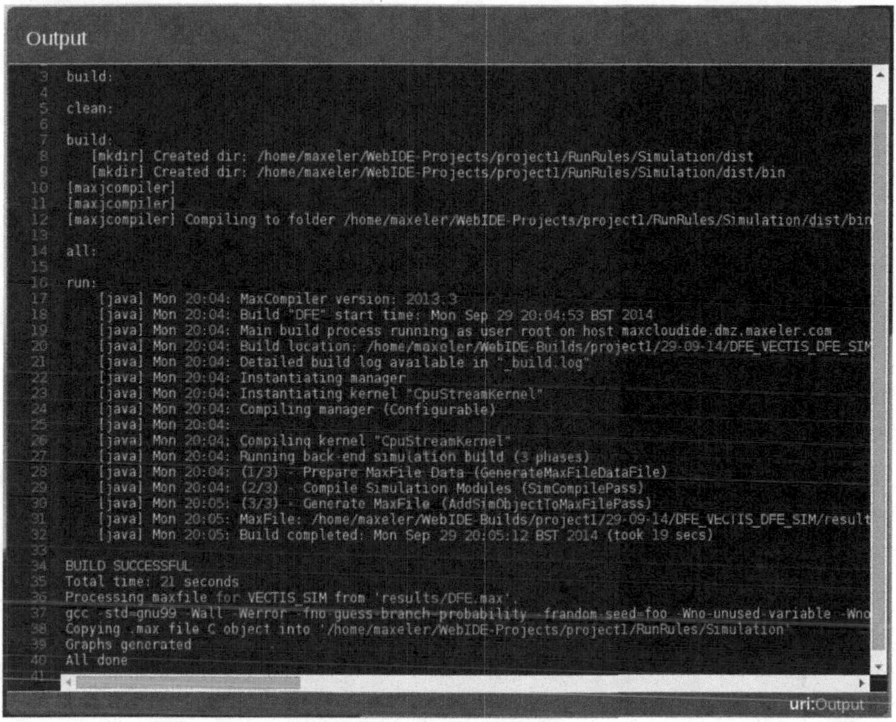

```
Output

 3  build:
 4
 5  clean:
 6
 7  build:
 8      [mkdir] Created dir: /home/maxeler/WebIDE-Projects/project1/RunRules/Simulation/dist
 9      [mkdir] Created dir: /home/maxeler/WebIDE-Projects/project1/RunRules/Simulation/dist/bin
10  [maxjcompiler]
11  [maxjcompiler]
12  [maxjcompiler] Compiling to folder /home/maxeler/WebIDE-Projects/project1/RunRules/Simulation/dist/bin
13
14  all:
15
16  run:
17      [java] Mon 20:04: MaxCompiler version: 2013.3
18      [java] Mon 20:04: Build "DFE" start time: Mon Sep 29 20:04:53 BST 2014
19      [java] Mon 20:04: Main build process running as user root on host maxcloudide.dmz.maxeler.com
20      [java] Mon 20:04: Build location: /home/maxeler/WebIDE-Builds/project1/29-09-14/DFE_VECTIS_DFE_SIM
21      [java] Mon 20:04: Detailed build log available in "_build.log"
22      [java] Mon 20:04: Instantiating manager
23      [java] Mon 20:04: Instantiating kernel "CpuStreamKernel"
24      [java] Mon 20:04: Compiling manager (Configurable)
25      [java] Mon 20:04:
26      [java] Mon 20:04: Compiling kernel "CpuStreamKernel"
27      [java] Mon 20:04: Running back-end simulation build (3 phases)
28      [java] Mon 20:04: (1/3) - Prepare MaxFile Data (GenerateMaxFileDataFile)
29      [java] Mon 20:04: (2/3) - Compile Simulation Modules (SimCompilePass)
30      [java] Mon 20:05: (3/3) - Generate MaxFile (AddSimObjectToMaxFilePass)
31      [java] Mon 20:05: MaxFile: /home/maxeler/WebIDE-Builds/project1/29-09-14/DFE_VECTIS_DFE_SIM/result
32      [java] Mon 20:05: Build completed: Mon Sep 29 20:05:12 BST 2014 (took 19 secs)
33
34  BUILD SUCCESSFUL
35  Total time: 21 seconds
36  Processing maxfile for VECTIS_SIM from 'results/DFE.max'.
37  gcc -std=gnu99 -Wall -Werror -fno-guess-branch-probability -frandom-seed=foo -Wno-unused-variable -Wno
38  Copying .max file C object into '/home/maxeler/WebIDE-Projects/project1/RunRules/Simulation'
39  Graphs generated
40  All done
41

                                                                              uri:Output
```

Fig. 4.12 Screenshot of Build Output component

4.15   Folder Explorer (Fig. 4.13)

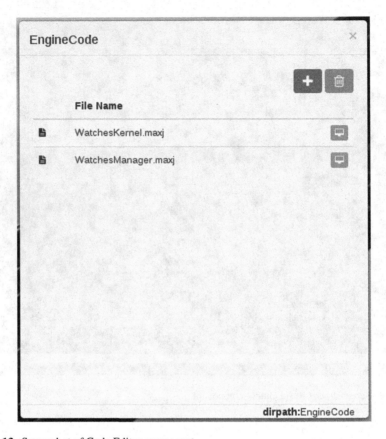

Fig. 4.13 Screenshot of Code Editor component

4.16   Image Viewer (Fig. 4.14)

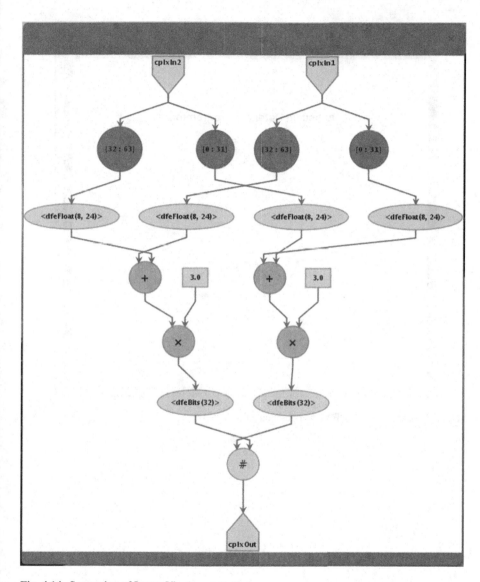

Fig. 4.14 Screenshot of Image Viewer component

4.17   PDF Viewer (Fig. 4.15)

Fig. 4.15 Screenshot of PDF Viewer component

4.18 CSV Viewer (Fig. 4.16)

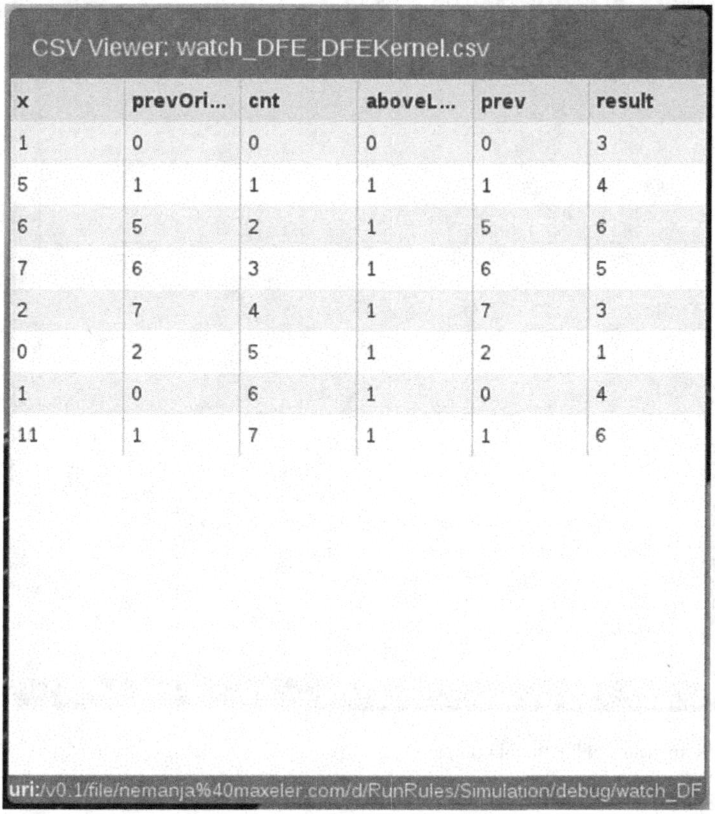

x	prevOri...	cnt	aboveL...	prev	result
1	0	0	0	0	3
5	1	1	1	1	4
6	5	2	1	5	6
7	6	3	1	6	5
2	7	4	1	7	3
0	2	5	1	2	1
1	0	6	1	0	4
11	1	7	1	1	6

CSV Viewer: watch_DFE_DFEKernel.csv

uri:/v0.1/file/nemanja%40maxeler.com/d/RunRules/Simulation/debug/watch_DF

Fig. 4.16 Screenshot of CSV Viewer component

4.19 Terminal

Fully functional terminal emulator has been built in WebIDE. You can use it to edit files using your favorite editor (Vim, Emacs, or Nano) and do many advanced operations that are you are not able to perform using WebIDE user interface like monitoring usage using top command (Fig. 4.17).

Fig. 4.17 Screenshot of Terminal component

4.20   WebIDE Change Log

```
+-----+------------+-------------------------------------------+
|Ver  | Date       |     Message              |
+-----+------------+-------------------------------------------+
|v0.1 | 06.Jun.2014 | Initial release.                          |
+-----+------------+-------------------------------------------+
|v0.2 | 07.Jun.2014 | Removed external dependencies (CDNs).     |
+-----+------------+-------------------------------------------+
|v0.3 | 28.Jul.2014 | + Added training examples;                |
|     |            | + Numerous small improvements.            |
+-----+------------+-------------------------------------------+
|v0.4 | 02.Sep.2014 | Added support for DFE builds.             |
+-----+------------+-------------------------------------------+
```

Epilogue

For the latest developments in the year 2015, the interested reader is referred to the website with a number of fully developed and fully explained examples tuned to both users and designers: The DataFlow Application Gallery (appgallery.maxeler.com).

© Springer International Publishing Switzerland 2015
V. Milutinović et al., *Guide to DataFlow Supercomputing*, Computer
Communications and Networks, DOI 10.1007/978-3-319-16229-4

Postscript

One of the coauthors of this book recently indicated at a conference that there is an analogy between DataFlow computing and lightning.

In both cases, it is the voltage difference that moves the relevant stuff, data in the case of DataFlow computing and electricity in the case of lightning.

This analogy, he said, could serve as an indication that, under specific conditions, the DataFlow approach could become many times faster than the control flow approach.

A similar conclusion could be generated if the research results of Richard Feynman are studied carefully. Remember, in the DataFlow approach, one writes a program to configure the hardware, and when data comes to the DataFlow accelerator input, driven by the voltage difference (between input and output), it (data) moves to the output, conditionally speaking, like lightning.

Based on the observations of Richard Feynman, theoretically, DataFlow could be many times faster, since control flow involves communications and DataFlow may not involve communications at all, if the compiler is smart enough, and if the topology of the underlying configurable hardware is closely corresponding to the topology of the DataFlow graph.

That is why it makes sense to refer to the DataFlow approach as the Feynman paradigm (contrary to the von Neumann paradigm used to describe the control flow approach).

© Springer International Publishing Switzerland 2015
V. Milutinović et al., *Guide to DataFlow Supercomputing*, Computer Communications and Networks, DOI 10.1007/978-3-319-16229-4

Index

© Springer International Publishing Switzerland 2015   127
V. Milutinović et al., *Guide to DataFlow Supercomputing*, Computer
Communications and Networks, DOI 10.1007/978-3-319-16229-4

Printed in the United States
By Bookmasters